Electromotive Force and Measurement in Several Systems

Electromotive Force and Measurement in Several Systems

Editor

Jens Omdahl

scitus academics

Electromotive Force and Measurement in Several Systems
Edited by **Jens Omdahl**

ISBN: 978-1-68117-297-2
Library of Congress Control Number: 2016937482

SCITUS Academics LLC,
www.scitusacademics.com
Box No. 4766, 616 Corporate Way,
Suite 2, Valley Cottage,
NY 10989

Printed in United States of America on Acid Free Paper

Preface

Electromotive force, also known as emf, is the voltage generated from any source of electrical energy, such as a battery or dynamo. The electromotive force is the maximum possible difference between two electrodes of the galvanic or voltaic cells. Energy is converted from one form to another in the generator or battery as the device works on the electrical charge transferred within it. One device terminal is loaded positively and the other is loaded negatively. The electric force is the work done on an electric charge unit or the energy obtained by it. Electromotive force is the feature of any energy source that can drive electric charge around a circuit. In this case, the term "force" does not mean mechanical force, measured in Newtons, but a potential or energy per load unit, measured in volts. The unit of electromotive force is the statvolt or one erg per electrostatic charge unit in the centimeter–gram –second system electrostatic units. When fluctuations in the magnetic field occur across the surface, emf is generated in nature. The shift of the Earth's magnetic field induces currents in the electric grid during a geomagnetic storm, as the magnetic field lines are moved around and cut across the conductor. This title focuses on various aspects of electromotive power theory and its applications in engineering science and industry.

Table of Contents

CHAPTER 1

Electromotive Force Generation with Hydrogen Release by Salt Water Flow under a Transverse Magnetic Field

Roberto De Luca

Dipartimento di Matematica e Informatica (DMI), Università degli Studi di Salerno, Fisciano, Italy

ABSTRACT

By considering an electrolyte solution in motion in a duct under a transverse magnetic field, we notice that a so called Faraday voltage arises because of the Lorentz force acting on anions and cations in the fluid. When salt water is considered, hydrogen production takes place at one of the electrodes if an electric current, generated by Faraday voltage, flows in an external circuit. The maximum amount of hydrogen production rate is calculated by basic electrochemical concepts.

KEYWORDS

Renewable Energies, Hydrogen Production, Faraday Voltage

1. INTRODUCTION

Aqueous ionic solutions and conducting metals show some similarities in their electrical transport properties. One of the simplest phenomenological laws of classical electrodynamics is Ohm's law of conduction [1], namely

$$\boldsymbol{J} = \sigma \boldsymbol{E} \tag{1}$$

where $\boldsymbol{J}$ is the current density flowing under the effect of the electric field $\boldsymbol{E}$, and s is the electrical conductivity of the material. A similar relation can be found in electrolyte solutions. In the latter case, the partial conductivities can be expressed in terms of the mobility m$_P$ of cations and m$_N$ of anions, so that:

$$\sigma_{P,N} = \rho_N e \mu_{P,N} \tag{2}$$

where ρ_N is the volume density of both types of ions and e is the electron charge. Mobility, on its turn, is defined as the proportionality constant linking the ions velocity $\boldsymbol{v}_{P,N}$ and the electric field $\boldsymbol{E}$, according to the following: $\boldsymbol{v}_{P,N} = \mu_{P,N}\boldsymbol{E}$. We recall that the total conductivity s is given by $\sigma_N + \sigma_P$.

It is well known that Lorentz force acts on charged particles moving in the presence of a magnetic field. Let us then consider anions and cations in an electrolyte solution flowing in a duct with velocity $\boldsymbol{v} = \boldsymbol{v}_E + \boldsymbol{v}'_{P,N}$, where $\boldsymbol{v}'_{P,N}$ is the velocity of the particles measured with respect to a reference frame moving with the aqueous solution. In the presence of a constant magnetic field $\boldsymbol{B}_0$, these ions are subject to a Lorentz force given by:

$$\boldsymbol{F}_{P,N} = \pm Ze\boldsymbol{v} \times \boldsymbol{B}_0 \tag{3}$$

where the plus and minus signs refer to cations and anions, respectively, both assumed to have ionic charge Ze in absolute value (Z thus being their valence). Because of Lorentz force, a voltage difference f_0, known in the literature as Faraday voltage [2,3], appears between the electrodes A and B of **Figure 1**. This voltage, similarly to what observed in the Hall's effect in metals [1], can be expressed as the product $B_0 v_A D$, when the electrolyte flows in a circular pipe of diameter D. Notice that in the present work boldface letters and corresponding plain text letters are used for vectors and their amplitudes, respectively. Geselowitz [4] has shown that this result is valid for any velocity profile, provided the flow is axially symmetric. It can also be shown that this result is valid, in particular, for a thin rectangular wire [5].

In the present work we propose an electromotive force (e.m.f.) generator consisting of a rectangular pipe with two lateral conducting plates enclosing salt water in motion under a magnetic field orthogonal to the flow velocity vector (see **Figure 1**). A similar system has already been proposed by Yamaguchi et al. [6] for a non-poison electro-conductive polymer and its mixture with magnetic fluid. We here also notice that, taking salt water solution as the electrolyte flowing in the rectangular pipe, hydrogen gas is formed at one of the two electrodes of the e.m.f. generator proposed. In the present work we calculate a general expression for the rate of production of hydrogen gas as a function of the flux rate of salt water flowing within the rectangular pipe. Experimental work is awaited to provide validation of the proposed analysis and to suggest further development of the topic of hydrogen production by salt water.

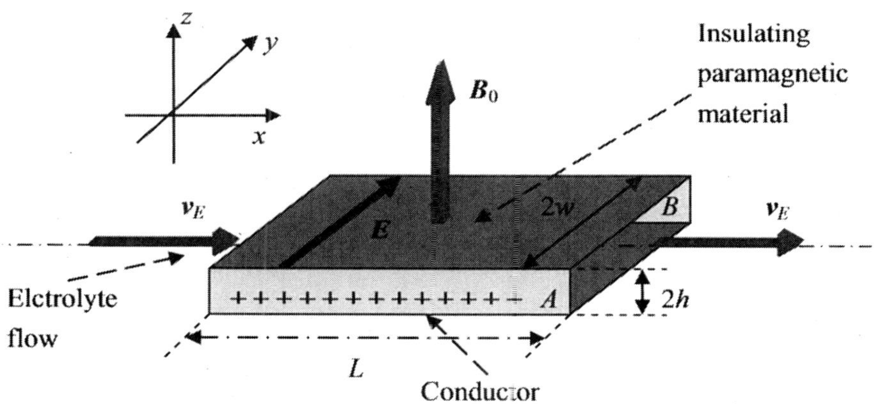

Figure 1. Schematic representation of the system: salt water flows through a rectangular pipe of section 4wh (2w is the width and 2h is the height) with velocity v in a volume of length L where a magnetic field B_0, orthogonal to V, is present. Because of Lorentz force, the charged ions in the solution accumulate in the vicinities of the lateral conducting plates, generating, at equilibrium, a constant electric field E.

2. FARADAY VOLTAGE

Let us consider the schematic view of the e.m.f. generator given in **Figure 1**. Take the magnetic field along the z-direction and the velocity $\boldsymbol{v}_E$ along the x-axis. Assume a velocity profile $v_E = v_E(y)$, and neglect, in the calculation of f_0, the amplitude of the drift velocities $\boldsymbol{v}'_{P,N}$ with respect to the average velocity of the fluid within the rectangular pipe $v_A = \frac{1}{2w}\int_{-w}^{w} v_E(y)\mathrm{d}y$. Under these hypotheses we may write the voltage drop f_0 across the two conductors as follows:

$$f_0 = 4v_A B_0 w \tag{4}$$

Notice that a factor of two appears in Equation (4), differently from what reported by Yamaguchi et al. [6], because of the presence of two ionic species in the electrolyte in our case. A detailed analysis of the dynamical problem is given in the Appendix, also to clarify how the low mobility coefficients for sodium and chlorine ions give drift velocities which are in amplitude much smaller than v_A.

By considering the equivalent circuit in **Figure 2**, where $r = w/\sigma hL$, $\sigma = \rho_N e(\mu_N + \mu_P)$ being the conductivity of the electrolyte solution, we may write:

$$P = 16\sigma v_A^2 B_0^2 K(1-K)hwL \tag{5}$$

where $K = R/(r+R)$ and a multiplicative factor of 4 is present when comparing the above result with that obtained by Yamaguchi et al.[6], because of Equation (4).

The above formula can be derived by setting $P = \Delta V^2/R$, where $\Delta V = K f_0$ is the voltage output of the device as shown in **Figure 2**. In order to estimate of the maximum power output (obtained for $K = 1/2$) in terms of the average water flow rate Q, for which $v_A = Q/4wh$, we write

$$P_{\max} = \frac{\sigma L}{4wh} Q^2 B_0^2 \tag{6}$$

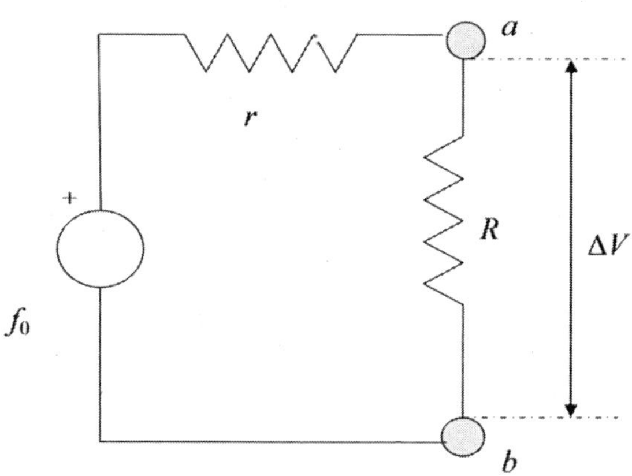

Figure 2. Equivalent circuit for an electric power generator with an open-circuit voltage f_0. When the internal resistance r is connected in series with an external load R, the output voltage is ΔV. Maximum power output $P_{\max} = f_0^2/4r$ is attained for $R = r$.

Therefore, for seawater at room temperature, by taking

$\sigma = 4.97\frac{1}{\Omega\mathrm{m}}$, $Q = 10.0\frac{l}{\mathrm{s}}$, $h = 0.05\ \mathrm{m}$, $w = 0.2\ \mathrm{m}$,

$L = 0.5\ \mathrm{m}$, and $B_0 = 2.0\ \mathrm{T}$, we have $P_{max} \approx 25\,\mathrm{mW}$ and $f_0 = 0.4\ \mathrm{V}$. Notice that the power output depends on the square of the average velocity v_A, in such a way that the device may be effective also in cases where the periodic motion of seawater could be used. The above calculation has been performed by assuming that the concentration of salt in water is sufficient to allow circulation of current in the system. However, as suggested by Wright and Van der Beken [2], and as calculated by the author [5], the average salinity of salt water is by far sufficient for current circulation in the system.

3. HYDROGEN PRODUCTION

In addition to the power output estimated in the previous section, we may notice that chemical reactions take place at the electrodes of the device in **Figure 1**. In fact, we may state that the current I can be sustained by oxidation of chlorine ions at one electrode (negatively charged), with production of chlorine gas, according to the following reaction [7]

$$2\mathrm{Cl}^- \rightarrow \mathrm{Cl}_2(g) + 2e^- \quad (7)$$

At the same time, reduction of water takes place at the opposite electrode according to the following additional reaction [7]

$$2Na^{+} + 2e^{-} + 2H_2O \rightarrow 2NaOH + H_2(g) \tag{8}$$

The reaction in (8) implies production of sodium hydroxide and hydrogen gas at the positively charged electrode. Therefore, also confiding on the experimental evidence in ref. [2], we may argue that it is possible to device an e.m.f. generator by letting salt water run in a pipe with average velocity $\boldsymbol{v}_A$ under an orthogonal magnetic field $\boldsymbol{B}_0$ and that, at the same time, it is also possible to produce hydrogen gas by this type of device.

The question is now the following: "How much hydrogen can be produced when a given amount of salt water flows inside the rectangular pipe in **Figure 1**?"

We start by assuming that, because of the reduction and oxidation reaction at the two electrodes, corresponding to the two lateral conducting plates of the device, it is possible to sustain a steady current flow in the system. However, the flow of electrons must be provided by a congruous number of ions moving along the pipe. Let us then also assume that only a fraction of charges in the volume $\Delta\tau = 4whL$ participate to the reductionoxidation processes specified in (7) and (8). When a stationary state sets in, the number of charges participating in the redox reactions must be such that a depletion of electron charges at the negative electrode is counterbalanced by the oxidation of chlorine ions, as seen in (7). Similarly, these charges must be also sufficient, in number, to compensate the excess of electrons at the other electrode by reduction of water as in (8). Contrarily to what assumed in a previous work [5], this aspect needs to be treated with great care. Indeed, the drift velocities in the y-direction v'_P and v'_N of the sodium and chlorine ions in salt water, respectively, can be seen to be much less than v_A in the Appendix. In this way, in the time $\Delta t = L/v_A$ only a small fraction of charges of those available in the volume $\tau_0 = 4hLw$ may be available for the redox reactions. In fact, if we consider the transverse current density given by (1), we may write $J_y = \sigma E$. Therefore, the available current I_0 is given by

$$I_0 = \frac{eN}{\Delta t} = 2hLJ_y = \frac{e\rho_N \tau_0}{w}\overline{v}_{P,N} \tag{9}$$

where $\overline{v}_{P,N} = \left(v'_P + v'_N\right)/2$. As it appears from Equation (9), N is much less than the total number of charged ions flowing in a time Δt in the region of length L of the rectangular pipe. We now need to compare the expression of I_0 with the maximum current $I_{max} = f_0/2r = 2v_A B_0 w/r$ circulating in the device when an external load of resistance $R = r$ is applied to it. Therefore, since $I_{max} = E\sigma hL$, we have $I_{max} = I_0/2$, which confirms the possibility of having ionic conduction according to the picture represented in the previous section.

The number of hydrogen moles n_{H} produced by the device in one second is then calculated by recognizing that one mole of H_2 is obtained when two moles of electrons are involved in the reactions at the electrons, in such a way that

$$\frac{\mathrm{d}n_{\mathrm{H}}}{\mathrm{d}t} = \frac{I_{max}}{2eN_{\mathrm{A}}} = \frac{\rho_N}{N_{\mathrm{A}}} hL\mathrm{v}_{\mathrm{A}} B_0 \left(\mu_N + \mu_P\right) \tag{10}$$

By now setting $v_A = Q/4wh$, the above expression can be recast in the following form

$$\frac{dn_H}{dt} = \frac{\rho_N}{N_A}\frac{L}{4w}QB_0\left(\mu_N + \mu_P\right) \tag{11}$$

which gives the rate of hydrogen production in a salt water solution under a constant magnetic field B_0 when the flux flow rate of the solution is Q. Typical values for seawater concentrations at 25°C can be seen to be

$n_{SW} = \frac{\rho_N}{N_A} = 0.43\frac{\text{mole}}{\text{l}} = 430\frac{\text{mole}}{\text{m}^3}$. Therefore, by taking again $Q = 10.0\frac{l}{\text{s}}$, $w = 0.2\ \text{m}$, $L = 0.5\ \text{m}$, and $B_0 = 2.0\ \text{T}$, we have $\frac{dn_H}{dt} = 0.64\times10^{-6}\frac{\text{mole}}{\text{s}}$.

4. CONCLUSIONS

In the present work it has been shown that an electromotive force generator can be realized by letting salt water run though a rectangular pipe under the influence of a transverse magnetic field. Because of Lorentz force on the moving sodium and chlorine ions in salt water, one can shown that an equilibrium electric field can be created by the accumulation of charged ions in the vicinities of the lateral conducting plates. By considering the problem in one dimension, it can be shown that the charge distribution varies linearly from one electrode to the other, if the velocity profile of the fluid in the pipe is parabolic [5]. Furthermore, it can be argued that the electric field across the electrodes can sustain a constant electromotive force f_0, which, on its turn, can generate an electric power output.

The device could be used as an alternative electric power source in situations where salt water is forced to run in large closed ducts, as it happens, for example, in some desalination plants. Experiments on small scale devices carried out by Wright and Van Der Beken [2] for a pipe with circular section do confirm the existence of a Faraday voltage across the device. In the present paper it is also predicted that, according to properties of ions in aqueous solutions, hydrogen gas can be produced at the positively charged lateral conducting plate. The rate of production of hydrogen gas is however small for not large values of salt water flow rate. Nevertheless, the present analysis can be seen as a starting point for further development of similar applications in the realm of alternative energy devices. In particular, provided experimental validation of the electro-chemical properties of the proposed system, one might envision applications where the increase in the concentration or the flow rate of salt water allows better hydrogen production rates and larger power output.

5. ACKNOWLEDGEMENTS

The author would like to thank A. Fedullo for useful discussions

REFERENCES

1 C. Kittel, "Introduction to Solid State Physics," John Whiley, New York, 1959.

2 J. J. Wright and S. Van Der Beken, "The Hall Effect in a Flowing Electrolyte," American Journal of Physics, Vol. 40, No. 2, February 1972, pp. 245-247.

3 H. S. T. Driver, "Comment on the Hall Effect in a Flowing Electrolyte," American Journal of Physics, Vol. 46, No. 12, December 1978, pp. 1275-1276.

4 D. B. Geselowitz, "Comment on the Hall Effect in a Flowing Electrolyte," American Journal of Physics, Vol. 40, No. 8, August 1972, pp. 1183-1184.

5 R. De Luca, "Lorentz Force on Sodium and Chlorine Ions in a Salt Water Solution Flow under a Transverse Magnetic Field," European Journal of Physics, Vol. 30, No. 3, May 2009, pp. 459-466.

6 H. Yamaguchi, X.-R. Zhang, S. Higashi and M. Li, "Study on Power Generation Using Electro-Conductive Polymer and Its Mixture with Magnetic Fluid," Journal of Magnetism and Magnetic Materials, Vol. 320, No. 7, 2008, pp. 1406-1411.

7 P. Mazzoldi, M. Nigro and C. Voci, "Fisica," EdiSES, 1998.

CHAPTER 2

Improved Sensorless Control of Interior Permanent Magnet Sensorless Motors Using an Active Damping Control Strategy

Younghoon Cho

Department of Electrical Engineering, Konkuk University, Seoul 05029, Korea; Tel.: +82-10-6207-0431; Fax: +82-447-9186

ABSTRACT

This paper proposes the active damping control strategy for position sensorless operation of an interior permanent magnet (IPM) motor. The proposed method is applied to both the current controller and the position estimator to control damping characteristics of the IPM drive system. By actively increasing the damping characteristics of the system with the proposed method, the current control and the position estimation loops become immune to parameter variation of the stator resistance which may degrade the accuracy of the position estimator. To analyze the accuracy of the position estimator with and without the proposed method, a small-signal analysis is carried out for low speed operation where the effect of the parameter variation is relatively large due to a low signal-to-noise ratio (SNR). Additionally, an open-loop voltage to angular velocity transfer function including the electrical and the mechanical parameters is investigated. Since no hardware modifications are necessary, the proposed method can be easily implemented just in software routines. Both the simulations and the experimental validations in which the proposed active damping control strategy is incorporated with the existing extended electromotive force (EMF)- based sensorless algorithm are provided to support the effectiveness of the proposed method.

KEYWORDS

active damping; interior permanent magnet motor; sensorless drive; extended electromotive force; position estimator; parameter variation

1. INTRODUCTION

Position sensors are necessary for high performance vector control of AC motors. However, popular position sensors such as optical encoders and resolvers are usually bulky, mechanically vulnerable, and may increase the implementation cost of the entire motor drive system. If the accuracy of the position sensor is unreliable, it may cause severe instability problems in the control loop. To overcome these drawbacks of position sensors, numerous control techniques where no position sensors are employed have been proposed [1]. The main purpose of sensorless control algorithms is to estimate the position and the speed of the motor without using a position or a speed sensor. In general, the realization of sensorless control can be classified into two categories: signal injection [2,3,4,5,6,7] and model-based methods [8,9,10,11,12]. In the signal injection methods, a high frequency voltage or current signal is directly utilized to estimate electrical position of the rotor by detecting inductance variation depending on rotor position and inductor saturation. These methods are generally suitable for low speed or zero speed operation due to additional losses and acoustic noise induced by injecting the voltage or current signal. The computational burden of implementing signal injection and filtering algorithms is an extra drawback.

The model-based methods mostly use electrical or electro-mechanical models of AC motors to estimate control parameters. Mostly, they extract the estimated states from back electromotive force (EMF) information. It means these methods cannot be directly applied under low speed or zero speed operation where back EMF information is rarely obtained. Many modern control schemes using a model reference adaptive system (MRAS), an adaptive observer, nonlinear control, and so on are proposed for sensorless AC motor drives [8,9,10,11,12]. In some cases, hybrid methods incorporating signal injection and model based methods are utilized to cover full operation range of the AC motor [13,14,15,16,17]. In many applications such as fans, compressors and blowers that normally have medium or high speed regions, the model-based methods are attractive [18,19,20].

Among many model-based sensorless methods, so-called extended EMF (EEMF)-based methods have been popularly adapted for IPM and synchronous reluctance motors where magnetic saliency exists [21,22,23,24]. However, the problem with these methods is that sensorless operation can fail due to parameter mismatches or variations, especially in low speed operation or mode transition [25]. In [26], online stator flux and resistance estimators are proposed to improve sensorless operation at low speed. A flux estimation algorithm that uses the phase current derivative without signal injection in low speed is presented in [27]. In [28], an online parameter estimation technique is employed to minimize the effects of parameter variations. In [29], the current measurement error and the inverter irregularities are analyzed, then a stator resistance estimator is proposed. The parameter identification and inverter error compensation algorithms based on the extended EMF method are presented in [30]. In [31,32], non-ideal characteristics of inverters such as output voltage phase delay and on-state resistance of the power devices are compensated to achieve performance improvement of the sensorless drive. However, the drawbacks of the aforementioned methods are that they are somewhat complicated to implement, and moreover, the performance of sensorless control is highly dependent on the accuracy of the position estimators and the compensation methods.

In this paper, an active damping control strategy is applied to improve the sensorless operation of the IPM motor. It is analyzed that low damping component in the entire control loop causes the failure of the sensorless operation under low speed or the mode transition between the open-loop acceleration and the sensorless mode. The proposed method artificially increases the damping component of the drive system so that the effects of parameter variations on the stator resistance are reduced. Therefore, stable sensorless operation in the low speed region where the signal-to-noise (SNR) is extremely low can be achieved, even with parameter mismatches between the physical system and the plant model. The small-signal analysis is taken to show the accuracy of the position estimator considering a practical parameter mismatch with and without the proposed method. From this, it is theoretically verified that the proposed method actively changes the damping component, and reduces the position estimation error. The open-loop voltage to angular velocity model is also derived to show the performance of the active damping control algorithm. Both the simulation and the experimental results for the IPM drive system show good agreement with the analyses, and the effectiveness of the proposed method are verified.

2. EXTENDED EMF BASED SENSORLESS CONTROL OF IPM MOTOR

In an IPM motor, the extended EMF E_{ex} is defined as [21,22]:

$$E_{ex} = \omega_r \left[(L_d - L_q)\, i_d + \lambda_f \right] - (L_d - L_q)\,(s i_q) \tag{1}$$

where s, ω_r, L_d, L_q, λ_f, i_d and i_q are the Laplace variable, the electrical angular velocity, the d-axis inductance, the q-axis inductance, the magnet flux linkage, and the d- and q-axes currents, respectively. By using Equation (1), the electrical model of IPM motors in the synchronous rotating frame is written as follows:

$$\begin{bmatrix} v_d \\ v_q \end{bmatrix} = \begin{bmatrix} R_s + sL_d & -\omega_r L_q \\ \omega_r L_q & R_s + sL_d \end{bmatrix} \begin{bmatrix} i_d \\ i_q \end{bmatrix} + \begin{bmatrix} 0 \\ E_{ex} \end{bmatrix} \tag{2}$$

where R_s, v_d, and v_q represents the stator resistance, the d-, and the q-axes voltages. It should be noted that L_q appears in both of d-

and q-axes cross-coupling terms, and L_d is used in non-cross-coupling terms unlike the typical electrical model of IPM motors. Let us define γ-δ coordinate as the estimated axes obtained by using a proper sensorless algorithm. Then, the electrical model of IPM motors oriented to γ-δ coordinate is written as:

$$\begin{bmatrix} v_\gamma \\ v_\delta \end{bmatrix} = \begin{bmatrix} R_s + sL_d & -\omega_r L_q \\ \omega_r L_q & R_s + sL_d \end{bmatrix} \begin{bmatrix} i_\gamma \\ i_\delta \end{bmatrix} + \begin{bmatrix} e_\gamma \\ e_\delta \end{bmatrix} \tag{3}$$

where i_γ and i_δ represent the estimated γ- and δ-axis currents in γ-δ coordinate, and e_γ and e_δ are the estimated EMF components in γ-δ coordinate. Here, e_γ and e_δ are defined as [22]:

$$\begin{bmatrix} e_\gamma \\ e_\delta \end{bmatrix} = E_{ex} \begin{bmatrix} -\sin(\theta_r - \hat{\theta}_r) \\ \cos(\theta_r - \hat{\theta}_r) \end{bmatrix} + (\hat{\omega}_r - \omega_r) L_d \begin{bmatrix} -i_\delta \\ i_\gamma \end{bmatrix} \tag{4}$$

where θ_r, θ^r, and ω^r are the actual electrical position, the estimated electrical position, and the estimated electrical angular velocity of the rotor.

Figure 1 shows the relationship between d-q and γ-δ coordinates. In sensorless operation, the second term in right-hand-side (RHS) in Equation (4) can be neglected by assuming L_d and the error between ω^r and ωr are small enough to be ignored after they are multiplied. By doing so, Equation (4) is simplified as Equation (5).

$$\begin{bmatrix} e_\gamma \\ e_\delta \end{bmatrix} \approx E_{ex} \begin{bmatrix} -\sin(\theta_r - \hat{\theta}_r) \\ \cos(\theta_r - \hat{\theta}_r) \end{bmatrix} \tag{5}$$

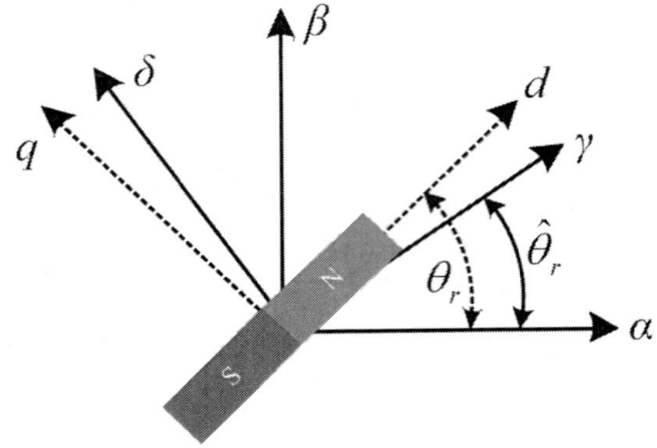

Figure 1. Relationship between *d-q* and *γ-δ* reference frames.

From Equation (5), the angle error θ_e between *d-q* and *γ-δ* coordinates is obtained as below:

$$\theta_e = \theta_r - \hat{\theta}_r = \tan^{-1}\left(-\frac{e_\gamma}{e_\delta}\right) \tag{6}$$

To calculate θ_e from Equation (6), Equation (3) is rearranged as follows:

$$\begin{aligned} e_\gamma &= v_\gamma + \omega_r L_q i_\delta - (R_s + sL_d)\, i_\gamma \\ e_\delta &= v_\delta - \omega_r L_q i_\gamma - (R_s + sL_d)\, i_\delta \end{aligned} \tag{7}$$

where v_γ and v_δ represent *γ*- and *δ*-axes voltages while i_γ and i_δ represent *γ*- and *δ*-axes currents. By substituting Equation (7) into

Equation (6), θ_e can be calculated. In practice, a low pass filter is multiplied on the equations in (7) as:

$$\begin{bmatrix} e_{\gamma LPF} \\ e_{\delta LPF} \end{bmatrix} = \frac{\omega_{ci}}{s+\omega_{ci}} \begin{bmatrix} e_{\gamma} \\ e_{\delta} \end{bmatrix} \tag{8}$$

where ω_{ci} is the radian expression of the cutoff frequency f_{ci} for the first order low pass filter. The low pass filter which was reported as a part of the observer in [22] helps to avoid direct implementation of the differential terms which may be vulnerable for noise components in feedback paths. Then, $e_{\gamma LPF}$ and $e_{\delta LPF}$ are rewritten as follows:

$$\begin{aligned} e_{\gamma LPF} &= \frac{\omega_{ci}}{s+\omega_{ci}}(v_{\gamma} + \omega_r L_q i_{\delta} - R_s i_{\gamma}) - \frac{\omega_{ci} s}{s+\omega_{ci}} L_d i_{\gamma} \\ e_{\delta LPF} &= \frac{\omega_{ci}}{s+\omega_{ci}}(v_{\delta} - \omega_r L_q i_{\gamma} - R_s i_{\delta}) - \frac{\omega_{ci} s}{s+\omega_{ci}} L_d i_{\delta} \end{aligned} \tag{9}$$

In fact, by introducing the low pass filter, the first term $\omega_{ci}/(s+\omega_{ci})$ in RHS of Equation (9) is analyzed as the first order Pseudo integrator with the cut-off angular frequency ω_{ci} while the second term $s\omega_{ci}/(s+\omega_{ci})$ operates as the first order Pseudo differentiator which may be the same form with a high pass filter. Compared to Equation (7), Equation (9) gives much more stable operation in steady state with slightly less, but acceptable, dynamic performance in back EMF estimation.

Figure 2 shows the structures of the position and speed estimator. The fundamental idea shown in Figure 2a is to estimateθ^r by tracking θ_e to be zero. In practice, θ_e is directly obtained from

Equation 6 as can be seen in Figure 2b, because ω_r or θ_r is not directly available. The estimator $G_{est}(s)$ in Figure 2 can be implemented with a simple proportional-integrator (PI) compensator. The detailed design procedure of $G_{est}(s)$ is explained in [22]. The electrical angle estimation algorithm is widely used in many applications due to its simple implementation and good estimation performance. However, there are mainly two problems with the algorithm. One is that the estimation performance is not very good with inaccurate system parameters. Another is that low speed operation may be hard or even impossible due to the difficulty of obtaining the extended EMF components in such a low signal-to-noise ratio (SNR) condition.

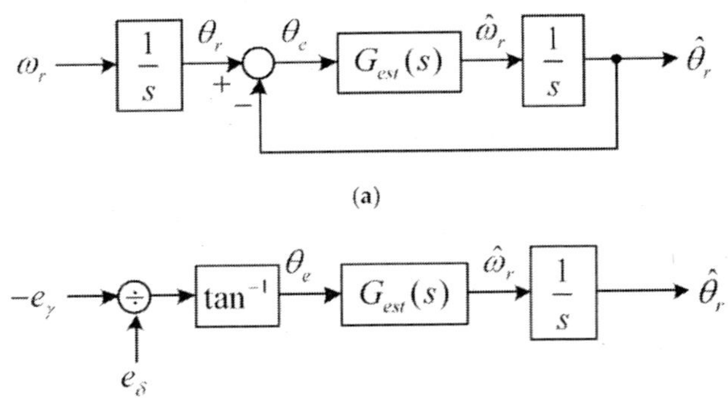

Figure 2. Structures of the position and speed estimator. (**a**) Basic concept. (**b**) Practical implementation.

3. SENSORLESS CONTROL WITH ACTIVE DAMPING CONTROL STRATEGY

The active damping technique used in this paper has been studied in several references [33,34,35,36] where the same concept may be called by different names. The basic concept of the active damping control is to artificially adjust the electrical damping of the motor drive system within the possible physical range by adding current feedback paths with a proper proportional gain. If the proportional gain is well selected, the stability and the dynamic performance of the control system can be improved [33,34,35,36]. In this paper, the active damping technique is combined with the back EMF estimation which is a part of the sensorless algorithm to achieve stable operation and superior dynamics of the IPM motor drive system.

3.1. Back EMF Estimation with Active Damping Control

The proposed back EMF estimation scheme using the active damping resistance R_{dp} is shown in Figure 3 where vγ_ff =ω^rLqiδ and vδ_ff=−ω^rLqiγ. Here, R_{dp} is added only to the estimated current feedback paths.

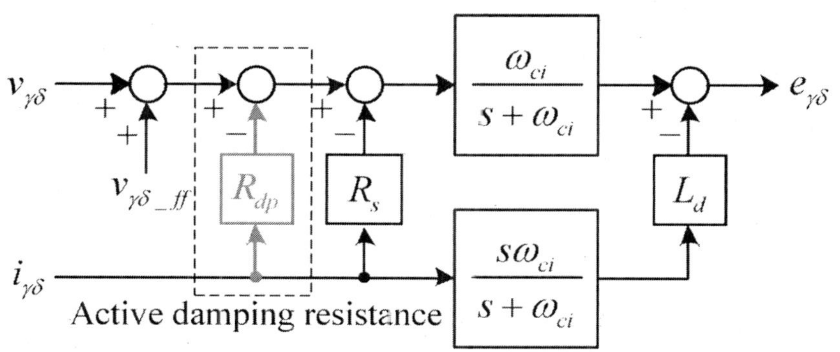

Figure 3. Back EMF estimation method using the active damping resistance.

The estimated back EMF components after passing the low pass filter can be written as follows.

$$\begin{aligned} e_{\gamma LPF} &= \frac{\omega_{ci}}{s+\omega_{ci}}\left(v_\gamma + v_{\gamma\_ff} - \left(R_s + R_{dp}\right) i_\gamma\right) - \frac{\omega_{ci}s}{s+\omega_{ci}} L_d i_\gamma \\ e_{\delta LPF} &= \frac{\omega_{ci}}{s+\omega_{ci}}\left(v_\delta + v_{\delta\_ff} - \left(R_s + R_{dp}\right) i_\delta\right) - \frac{\omega_{ci}s}{s+\omega_{ci}} L_d i_\delta \end{aligned} \tag{10}$$

Note that Equation (10) has the same form as Equation (9) except R_{dp} is added to R_s. The equations in (10) imply that the tendency of the back EMF and angle estimation characteristics can be adjusted by R_{dp}. In order to see how R_{dp} affects the estimation performance, the small signal analysis for θ_e will be discussed hereafter. For convenience of the analysis, the first order Pseudo integrator and differentiator are neglected. Then, i_δ can be rearranged as in Equation (11):

$$i_\delta = \frac{v_\delta - \omega_r L_q i_\gamma - e_\delta}{L_d s + (R_s + R_{dp})} \tag{11}$$

By substituting Equation (11) into e_γ, we have:

$$e_\gamma = v_\gamma + \omega_r L_q \left(\frac{v_\delta - \omega_r L_q i_\gamma - e_\delta}{L_d s + (R_s + R_{dp})}\right) - \left((R_s + R_{dp}) + sL_d\right) i_\gamma \tag{12}$$

If the operational speed and the inertia of the IPM motor are low and high enough, it is possible to assume that the differentiation of the mechanical speed is zero. After that, e_γ can be simplified as in Equation (13) by considering the perturbation:

$$\tilde{e}_\gamma = -\frac{\omega_r L_q \tilde{e}_\delta}{L_d s + (R_s + R_{dp})} \tag{13}$$

Equation (13) can be rearranged as below:

$$-\frac{\tilde{e}_\gamma}{\tilde{e}_\delta} = \frac{\omega_r L_q}{L_d s + (R_s + R_{dp})} \tag{14}$$

By substituting Equation (14) into Equation (6), the small signal model of the electrical angle error θ˜e is obtained as follows:

$$\tilde{\theta}_e = \tan^{-1}\left(-\frac{\tilde{e}_\gamma}{\hat{e}_\delta}\right) = \tan^{-1}\left(\frac{\omega_r L_q}{L_d s + (R_s + R_{dp})}\right) \tag{15}$$

If θ˜e changes within a very limited range, e.g., −0.2 rad $< \tilde{\theta}_e < 0.2$ rad, (15) can be simplified even more as below:

$$\tilde{\theta}_e = \tan^{-1}\left(\frac{\omega_r L_q}{L_d s + (R_s + R_{dp})}\right) \approx \frac{\omega_r L_q}{L_d s + (R_s + R_{dp})} \tag{16}$$

Figure 4 compares the frequency responses of θ˜e at different R_{dp} values where L_d = 2.51 mH, L_q = 6.94 mH, R_s = 0.09 Ω, and ω_r = 18.85 rad/s. As can be seen in the figure, the magnitude of $\tilde{\theta}_e$ is getting lower as R_{dp} increasing. If all parameters and variables are constant in (16) except R_{dp}, the gain of the transfer function is reduced as R_{dp} increases. This implies that even if the back EMF estimation is not very stable, the electrical angle error is smaller with larger R_{dp}. In other words, the angle estimation is better with R_{dp} in transients and some conditions where the back EMF estimation is oscillatory.

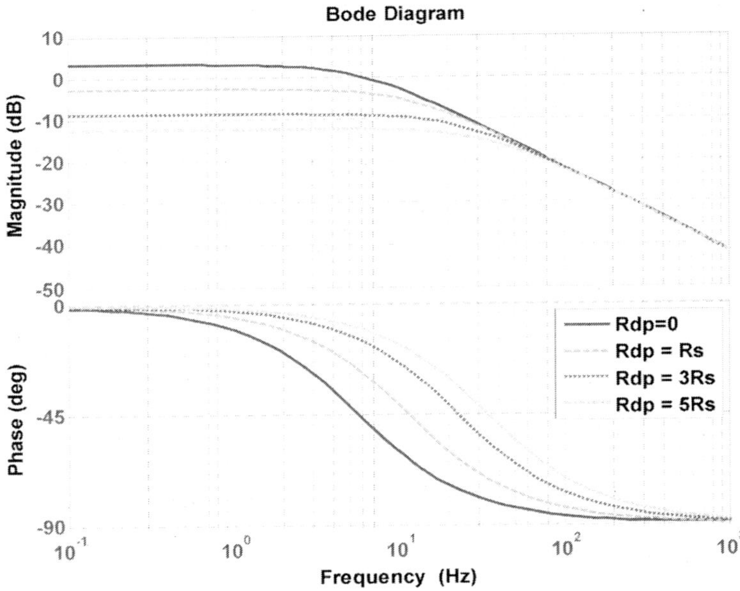

Figure 4. Comparison of the frequency responses of θ˜e at different R_{dp}.

The trajectories of the position error variation depending on the ratio of the BEMF variation are plotted in Figure 5. As illustrated in the figure, the variation of the position error becomes smaller as R_{dp} is increased. This trend also matches the analysis taken previously.

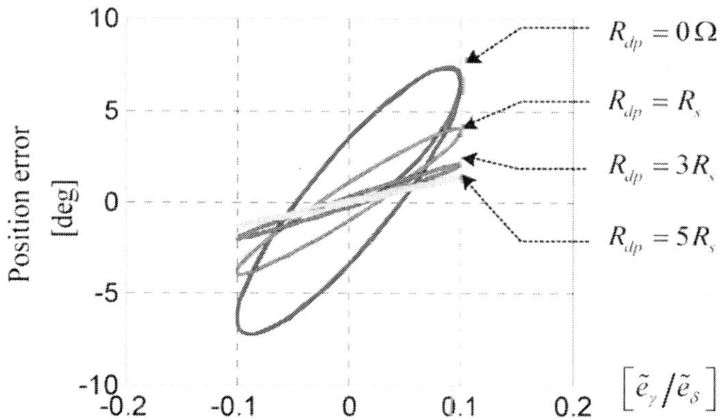

Figure 5. Trajectories of the position error variation.

3.2. Current Control Strategy with Active Damping Control

If R_{dp} is used for the back EMF estimation, it is better to use R_{dp} even in the current controller to reduce the model mismatch between the actual and the estimated control system as shown in Figure 6.

Figure 7 shows the current controller implementation with R_{dp}. Note that the current controller consists of a PI controller as a feedback controller and a decoupling terms v_{dq_ff} as follows:

$$\begin{bmatrix} v_{d\_ff} \\ v_{q\_ff} \end{bmatrix} = \begin{bmatrix} 0 & -\omega_r \\ \omega_r & 0 \end{bmatrix} \begin{bmatrix} L_d & 0 \\ 0 & L_q \end{bmatrix} \begin{bmatrix} i_d \\ i_q \end{bmatrix} \tag{17}$$

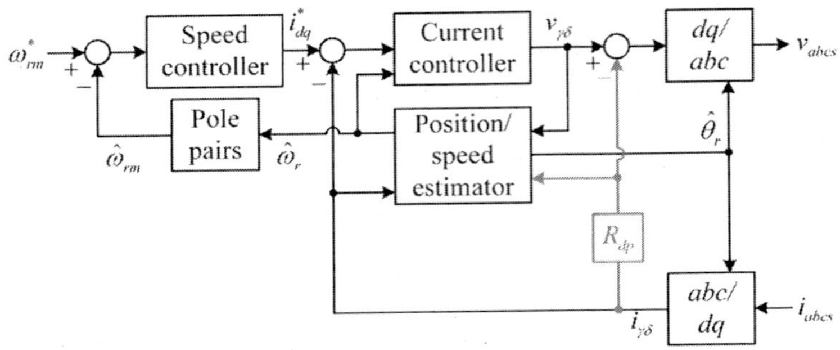

Figure 6. Sensorless control algorithm with the active damping control strategy.

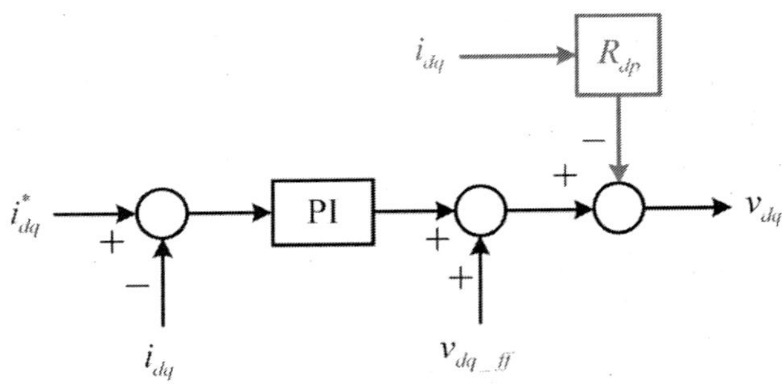

Figure 7. Current controller implementation with R_{dp}.

The gain of the PI controllers is selected as:

$$K_{pd} = \omega_c L_d K_{pq} = \omega_c L_q$$
$$K_{id} = K_{iq} = \omega_c \left(R_s + R_{dp} \right) \tag{18}$$

where K_{pd}, K_{pq}, K_{id}, K_{iq}, and ω_c represent d-, q-axes proportional gains, d-, q-axes integral gain, and desired bandwidth of the current controller, respectively. Note that R_{dp} is included to the integral gain. By selecting the control gain this way and using the feed-forward terms, the overall characteristics of the entire control system can be emulated as a simple low pass filter whose cutoff frequency is the same to the bandwidth of the control system, as expressed in Equation (19).

$$\frac{i_{dq}(s)}{i^*_{dq}(s)} \simeq \frac{\omega_c}{s+\omega_c} \tag{19}$$

This is so-called the technical optimization of current controllers [37]. Equation (19) implies that the current control bandwidth is identical to ω_c, and the phase margin is 90°.

To examine the effect of R_{dp} in the current control loop, a simple voltage to mechanical speed model will be derived hereafter. Let us consider the torque equation of the IPM motor as below:

$$T_e = \frac{3}{2}\frac{P}{2}\left(\lambda_f i_q + (L_d - L_q)\, i_d i_q\right) \tag{20}$$

where P is the number of poles. In order to simplify the analysis, i_d is assumed to be zero, then we have:

$$T_e = \frac{3}{2}\frac{P}{2}\lambda_f i_q = K_T i_q\,, \quad \text{where } K_T = \frac{3}{4}K_E \text{ and } K_E = P\lambda_f \tag{21}$$

The block diagram of q-axis voltage-to-mechanical speed is shown in Figure 8, whose parameters are denoted in Table 1. In order to see the effect of R_{dp} only, the feedback controllers are excluded in this analysis. From Figure 8, the transfer function of v_q- to -ω_{rm} is given as Equation (22):

$$G_{v\omega}(s) = \frac{\omega_{rm}}{v_q} = \frac{K_T}{\left(sL_q + \left(R_s + R_{dp}\right)\right)(J_m s + B_m) + K_T K_E} \quad (22)$$

where J_m and B_m are the inertia moment and the friction coefficients of the mechanical system. Figure 9 compares the frequency response of Equation (22) with different R_{dp} values. The highest resonance occurs with no R_{dp}. The resonance means that ω_{rm} is oscillatory when v_q which is the output of the feedback controller changes quickly in cases such as a step load variation. If R_{dp} is increased, the resonance is getting damped. When R_{dp} is selected to be $10R_s$, the resonance is almost nothing so that the frequency response of the transfer function is similar to a first order low pass filter. In this case, ω_{rm} may not fluctuate that much even in the step load variation. Accordingly, it is supposed that the use of R_{dp} definitely helps to stabilize the speed control loop of the motor drive system.

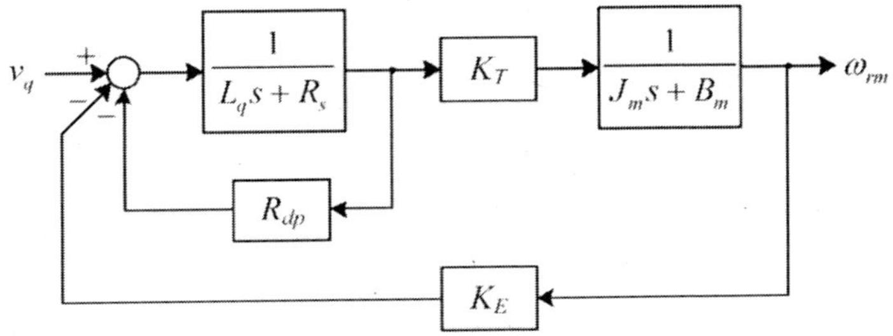

Figure 8. Block diagram to examine the effect of R_{dp}.

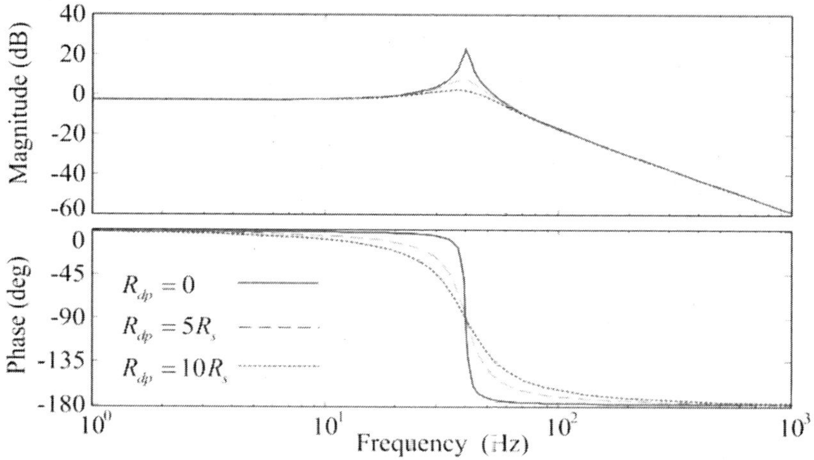

Figure 9. Frequency response of $G_{v\omega}(s)$ under different R_{dp} conditions.

Table 1. Motor parameters.

Parameters	Values
Number of poles (P)	6
Magnet flux linkage (λ_f)	0.235 V/(rad/s)
d-axis inductance (L_d)	2.51 mH
q-axis inductance (L_q)	6.94 mH
Stator resistance (R_s)	0.09 Ω
Rotor inertia (J_m)	0.003334 kg· m^2
Friction constant (B_m)	0.425 × 10^{-3}· m^2/s
Rated speed	1600 rev/min
Rated torque	65 Nm

4. SIMULATION STUDY

Simulations have been carried out to examine the performance of the proposed method on the sensorless control. The parameters in the simulations are the same with Table 1. The speed reference profile and the operation mode for the simulations are shown in Figure 10. To start the IPM motor, the well-known open-loop synchronous acceleration method [19,20] is applied. The EEMF based sensorless algorithm is also utilized to estimate the position and the speed of the IPM motor. The mode transition from the open-loop synchronous acceleration to the sensorless control occurs at 60 RPM.

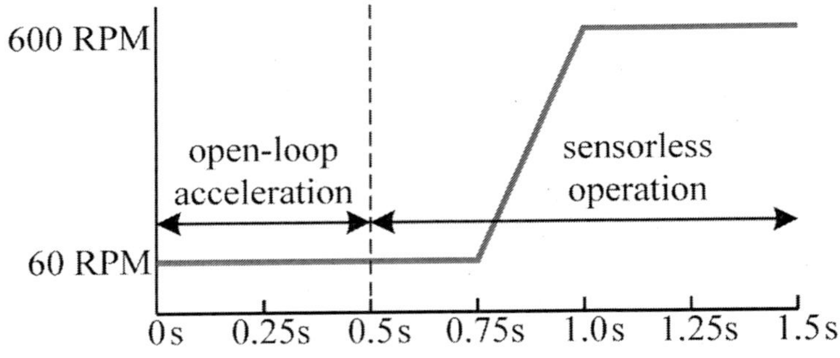

Figure 10. The profile of the speed reference and the operation modes.

Figure 11 shows the simulation results where the parameter mismatches in the current controller and the position and the speed estimator are intentionally inserted. At the start, 10% of the rated torque is applied. After 0.5 s, the operation mode is switched from the open-loop synchronous acceleration to the sensorless mode. After the speed of the motor is increased to 500 RPM, 30% of the rated torque is applied at t = 1.25 s. In each figure, the red dashed line, the green dotted line, and the blue solid line represent the cases when R_{dp} equals to zero, $3\hat{R}_s$, and $5\hat{R}_s$.

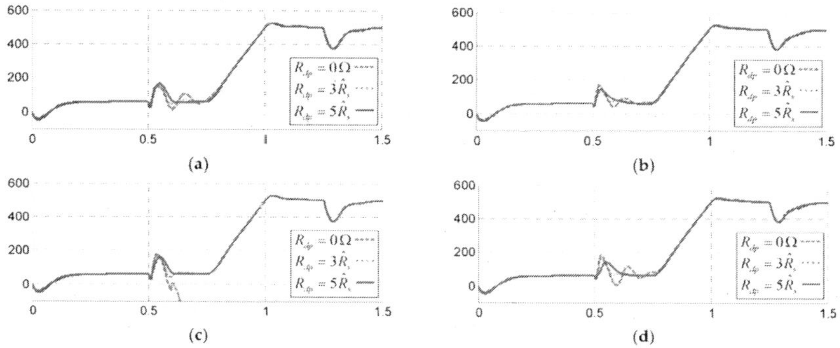

Figure 11. Simulation results. (**a**) R^s=0.75Rs, L^dq=0.75Ldq, and λ^f=1.15λf; (**b**) R^s=0.75Rs, L^dq=1.25Ldq, and λ^f=1.15λf; (**c**) R^s=1.25Rs, L^dq=0.75Ldq, and λ^f=1.15λf; (**d**) R^s=1.25Rs, L^dq=1.25Ldq, and λ^f=1.15λf.

In each case in Figure 11, the responses without R_{dp} show more fluctuation than the cases with R_{dp}, where the response is much more damped at the mode transition from the open-loop synchronous acceleration to the sensorless operation. This is especially apparent in Figure 11c where the system failed to switch the operation mode to the sensorless control. These results correspond to the analysis in the previous sections, and show that the mode transition is made more stable by using R_{dp}. This implies that R_{dp} helps the stabilization of the IPM motor drive system undergoing the mode transition in the low speed region. Once the motor speed is over 500 RPM, the responses of a 30% load step condition are similar in all of the cases. From this result, it can be inferred that the utilized R_{dp} does not affect the normal sensorless operation where the motor speed is high enough to overcome low SNR on the estimated EMF components. However, it should be noted that the feedback current multiplied by R_{dp} is not a real-time current, but a sampled current, which has a zero-order-hold (ZOH) effect. Hence, the feedback current may not be the same with an ideal case, because the high frequency components are delayed in the feedback. The limitation of the output voltage is also another issue. If the feedback current multiplied by $R_{dp}$is too high, the output voltage reference may be over the physical limitation of the output voltage. Therefore, too high value of R_{dp} may cause other instability issues in the system due to the ZOH effect and the output voltage limitation.

5. EXPERIMENTS

Experiments have been performed to show the effectiveness of the proposed method. Figure 12 shows the experimental configuration.

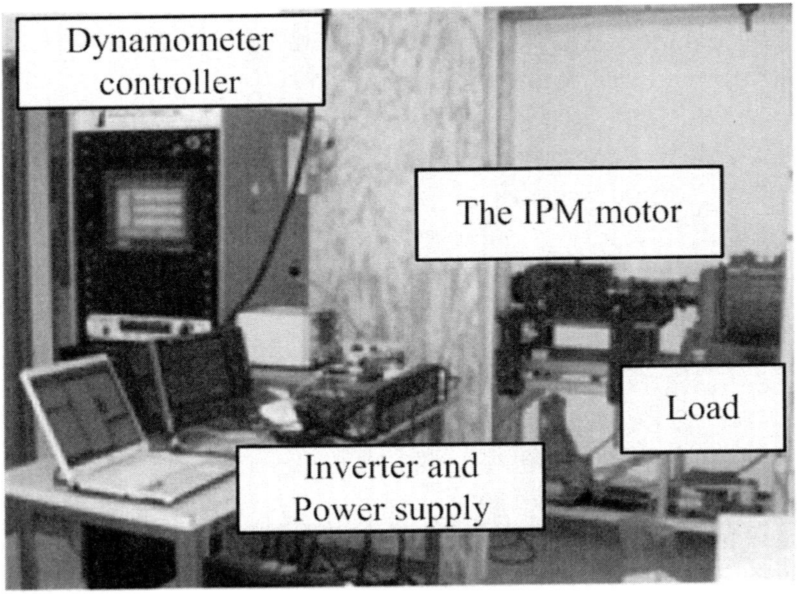

Figure 12. Experimental configuration.

The dc-link voltage is selected as 300V and it is fed by a dc power supply. The same motor parameters described in Table 1 are utilized for the experiments. The specifications of control parameters are shown in Table 2.

Table 2. Control parameters.

Parameters	Values
Switching frequency	10 kHz
Sampling frequency	10 kHz
Current control bandwidth	300 Hz
Position estimator bandwidth	100 Hz
Crossover frequency of the Pseudo integrator and differentiator	100 Hz

The proposed algorithm is implemented with a custom made digital control board in which a 32-bit digital signal processor is employed. The control board includes a 4-channel digital-to-analog converter (DAC) to monitor the internal variables in the software routines. A 13-bit external encoder is connected to the IPM motor to compare the real and the estimated position of the rotor. A Magtrol dynamometer is utilized to apply loads and to measure the motor torque. Although the testing motor is an IPM motor, the *d*-axis current reference is selected as zero to simplify confirming the validation of the proposed algorithm where a maximum torque per ampere is not necessary. However, the operational principle should not be different even in flux weakening region. In the experiments, the magnet flux linkage, the stator resistance, the *d*- and the *q*-axes inductances in the controller is selected as $0.9\lambda_f$, $0.75R_s$, $1.15L_d$, and $1.15L_q$ to artificially emulate the parameter mismatch condition. The active damping component R_{dp} is selected as $5\hat{R}_s$. In Figure 13, the sensorless operation with 10% of the rated torque is shown. Figure 13a shows the mode transition from the open-loop synchronous acceleration to the sensorless operation. The speed at the transition point is 120 RPM. As can be seen in the figure, the operation mode is smoothly switched. Figure 13b shows the steady state operation of the sensorless control, which is very stable. In both Figure 13a,b the oscilloscope channels 3 and channel 4 represent the actual electrical position $\theta_r$measured by the 13-bit encoder and the estimated position θ^$_r$. Channel 1 and channel 2 indicate the estimated EMFs e_γ ande_δ. The values of e_γ and e_δ are evaluated as 0V and 8.86V. These numbers agree with the results obtained from (3) when no errors occur. Hence, the position and the EMFs estimation are smooth, even with R_{dp}.

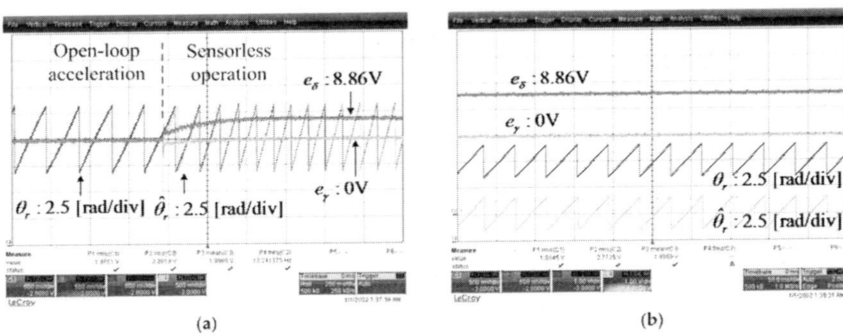

Figure 13. Test results of the sensorless operation: (**a**) Mode transition; (**b**) Steady state.

Figure 14 shows the test results of a deceleration where the speed reference is changed from 500 RPM to 60 RPM in ramp. Here, 30% of the rated torque is applied as the load. In both cases, the estimated EMF value and the speed control performance are within the normal range at the beginning. However, as seen in Figure 14a, the position and the EMF estimation fail and the speed is not regulated as the operating speed decreases without Rdp. This is due to the low SNR on the iterated values in the controller, the parameter mismatches, and the low system damping characteristic. Meanwhile, inFigure 14b, the stable sensorless operation is achieved with Rdp after slight transients at the beginning of 60 RPM operation under the same test condition.

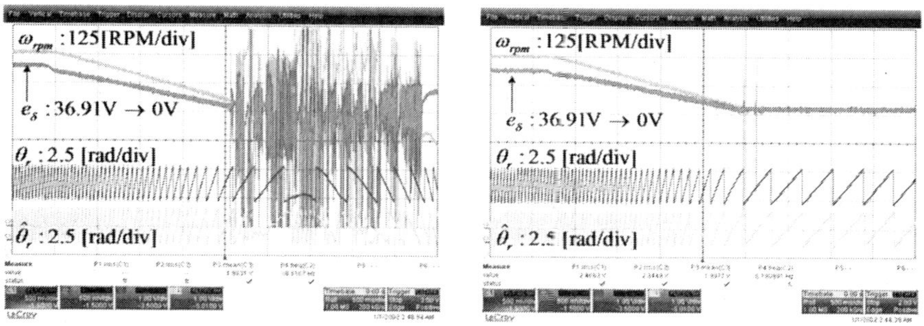

(a) (b)

Figure 14. Deceleration test results: (**a**) Rdp=0Ω (**b**) Rdp=5R^s.

Figure 15 compares the trajectories of the position error θ_e which depends on the estimated EMF ratio e_γ/e_δ. The operating speed is 90 RPM and no load torque is applied. As analyzed in the previous section, the dispersion of the error trajectory with R_{dp} is narrower than the case without R_{dp}. It means the performance of the position estimation is accurate with the active damping resistance.

Figure 16 compares the position and the EMF estimation performance without and with R_{dp}. The same operating condition as in Figure 15 is assumed. As shown in the figure, the peak to peak fluctuation of the estimated EMF and the position are evaluated as about 20V and 0.35 rad before applying R_{dp}. However, the magnitudes are decreased to 3V and 0.18 rad with R_{dp}. The ripple components in the estimated EMF and the position are also reduced.

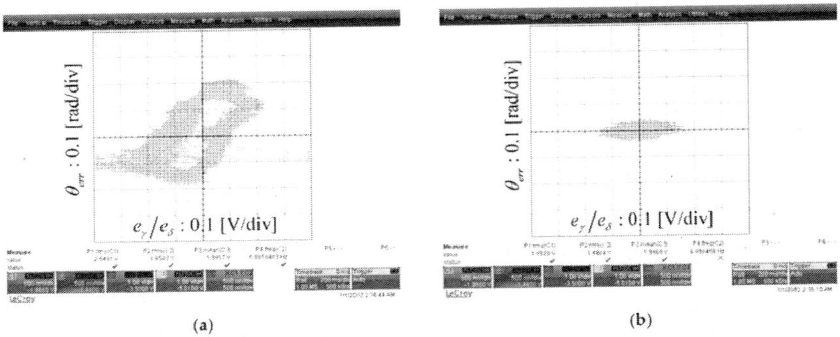

Figure 15. Comparison of the trajectories of θ_e: (**a**) Rdp= 0Ω (**b**) Rdp=5R^s.

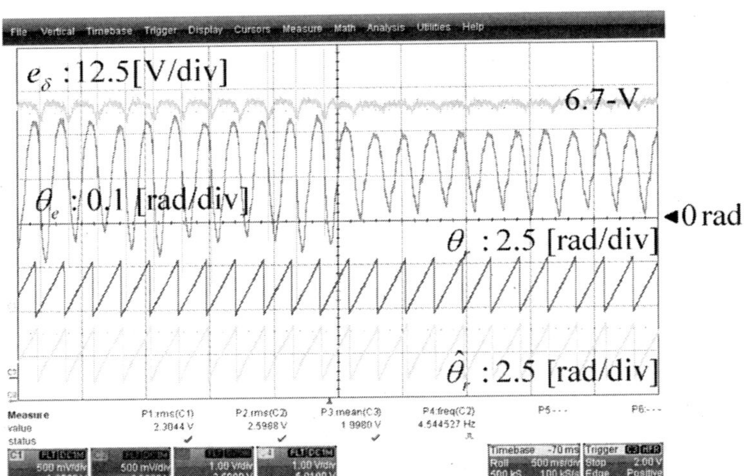

Figure 16. The position and the EMF estimation performance.

Figure 17 illustrates the test results on the torque and the speed response. The speed reference is 60 RPM and the applied load torque is 5% of the rated torque. In Figure 17a, the magnitude of the torque ripple is reduced from 3Nm to 1Nm after applying R_{dp}. The range of the speed fluctuation is also limited from 40 RPM to 10 RPM with R_{dp}.

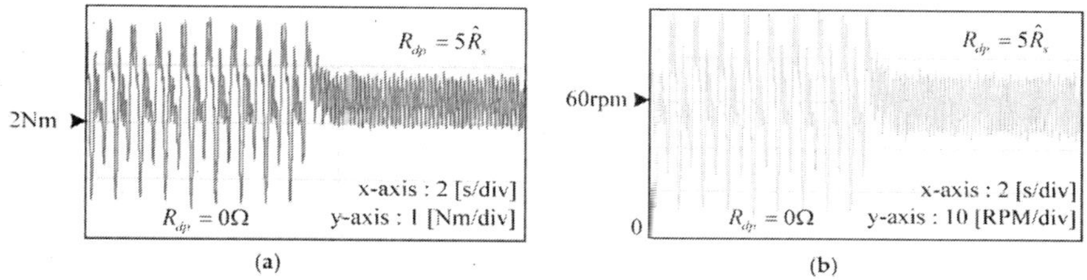

Figure 17. Torque and speed response without and with R_{dp}. (**a**) Torque response; (**b**) Speed response.

Figure 18 compares the step load test performance. 50% of the rated torque is applied using the Magtrol dynamometer. As can be seen in the comparison, if R_{dp} is not utilized, the speed regulation fails when the load is applied. With R_{dp}, although the minimum speed reaches 10 RPM, the speed regulation is recovered in steady state. In the experimental results, by using the proposed method, it is confirmed that the EMF and the position estimation performance and the stability of the system are improved even with the parameter mismatches.

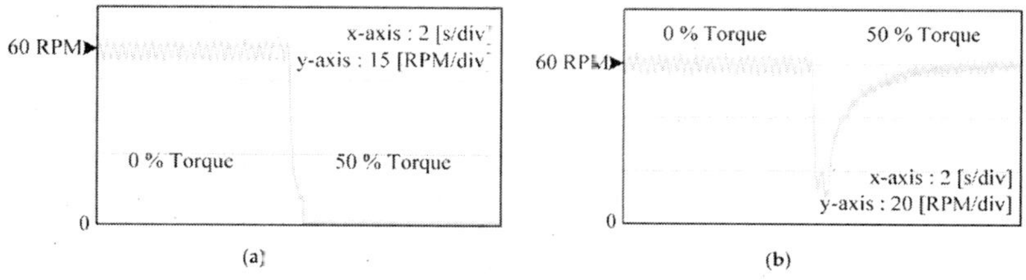

Figure 18. Speed response (**a**) without R_{dp} (**b**) with R_{dp}.

6. CONCLUSIONS

This paper proposed an active damping control strategy for the sensorless control of an IPM motor. The proposed method reduces the effects of the parameter mismatches, and increases the damping characteristics of the overall control system. Accordingly, the stability of the control system is improved. The small signal analysis for the position error and the frequency response of the open-loop voltage-to-speed transfer function are performed to evaluate the stability of the system using the proposed method. The proposed algorithm is adapted to the EEMF based sensorless method. Both the simulations and the experiments agree that the EMF and the position estimation performance are improved, and the torque and the speed ripple are reduced.

ACKNOWLEDGMENTS

All sources of funding of the study should be disclosed. Please clearly indicate grants that you have received in support of your research work. Clearly state if you received funds for covering the costs to publish in open access.

CONFLICTS OF INTEREST

The author declares no conflict of interest.

REFERENCES

1 Acarnley, P.P.; Watson, J.F. Review of position-sensorless operation of brushless permanent-magnet machines. *IEEE Trans. Ind. Electron.* **2006**, *53*, 352–362.

2 Ha, J.I.; Ide, K.; Sawa, T.; Sul, S.K. Sensorless rotor position estimation of an interior permanent-magnet motor from initial states. *IEEE Trans. Ind. Appl.* **2003**, *39*, 761–767.

3 Shi, J.L.; Liu, T.H.; Chang, Y.C. Position control of an interior permanent-magnet synchronous motor without using a shaft position sensor. *IEEE Trans. Ind. Electron.* **2007**, *54*, 1989–2000.

4 Kasa, N.; Watanabe, H. A mechanical sensorless control system for salient-pole brushless DC motor with autocalibration of estimated position angles. *IEEE Trans. Ind. Electron.* **2000**, *47*, 389–395.

5 Raca, D.; Garcia, P.; Reigosa, D.D.; Briz, F.; Lorenz, R.D. Carrier-signal selection for sensorless control of PM synchronous machines at zero and very low speeds. *IEEE Trans. Ind. Appl.* **2010**, *46*, 167–178.

6 Reigosa, D.D.; Garcia, P.; Raca, D.; Briz, F.; Lorenz, R.D. Measurement and adaptive decoupling of cross-saturation effects and secondary saliencies in sensorless controlled IPM synchronous machines. *IEEE Trans. Ind. Appl.* **2008**, *44*, 1758–1767.

7 Yoon, Y.D.; Sul, S.K.; Morimoto, S.; Ide, K. High bandwidth sensorless algorithm for AC machines based on square-wave type voltage injection. In Proceedings of the 2009 IEEE Energy Conversion Congress and Exposition, San Jose, CA, USA, 20–24 September 2009; pp. 2123–2130.

8 Hajian, M.; Soltani, J.; Markadeh, G.A.; Hosseinnia, S. Adaptive nonlinear direct torque control of sensorless IM drives with efficiency optimization. *IEEE Trans. Ind. Electron.* **2010**, *57*, 975–985.

9 Kim, M.; Sul, S.K. An enhanced sensorless control method for PMSM in rapid accelerating operation. In Proceedings of the 2010 International Power Electronics Conference (IPEC), Sapporo, Japan, 21–24 June 2010; pp. 2249–2253.

10 Matsui, N. Sensorless PM brushless DC motor drives. *IEEE Trans. Ind. Electron.* **1996**, *43*, 300–308.

11 Orlowska-Kowalska, T.; Dybkowski, M. Stator-current-based MRAS estimator for a wide range speed-sensorless induction-motor drive. *IEEE Trans. Ind. Electron.* **2010**, *57*, 1296–1308.

12 Piippo, A.; Hinkkanen, M.; Luomi, J. Analysis of an adaptive observer for sensorless control of interior permanent magnet synchronous motors. *IEEE Trans. Ind. Electron.* **2008**, *55*, 570–576.

13 Foo, G.; Rahman, M.F. Sensorless sliding-mode MTPA control of an IPM synchronous motor drive using a sliding-mode observer and HF signal injection. *IEEE Trans. Ind. Electron.* **2010**, *57*, 1270–1278.

14 Mizutani, R.; Takeshita, T.; Matsui, N. Current model-based sensorless drives of salient-pole PMSM at low speed and standstill. *IEEE Trans. Ind. Appl.* **1998**, *34*, 841–846.

15 Sayeef, S.; Foo, G.; Rahman, M.F. Rotor position and speed estimation of a variable structure direct-torque-controlled IPM synchronous motor drive at very low speeds including standstill. *IEEE Trans. Ind. Electron.* **2010**, *57*, 3715–3723.

16 Foo, G.; Sayeef, G.; Rahman, M.F. Low-speed and standstill operation of a sensorless direct torque and flux controlled IPM synchronous motor drive. *IEEE Trans. Energy Convers.* **2010**, *25*, 25–33.

17 Boldea, I.; Paicu, M.C.; Andreescu, G.D.; Blaabjerg, F. Active flux DTFC-SVM sensorless control of IPMSM. *IEEE Trans. Energy Convers.* **2009**, *24*, 314–322.

18 Raggl, K.; Warberger, B.; Nussbaumer, T.; Burger, S.; Kolar, J.W. Robust angle-sensorless control of a PMSM bearingless pump. *IEEE Trans. Ind. Electron.* **2009**, *56*, 2076–2085.

19 Rho, M.S.; Kim, S.M. Development of robust starting system using sensorless vector drive for a microturbine. *IEEE Trans. Ind. Electron.* **2010**, *57*, 1063–1073.

20 Yim, J.S.; Lee, W.J.; Sul, S.K.; Yang, H.S.; Kim, J.T. Sensorless vector control of super high speed turbo compressor. In Proceedings of the Twentieth Annual IEEE Applied Power Electronics Conference and Exposition, Austin, TX, USA, 6–10 March 2005; pp. 950–953.

21 Zhiqian, C.; Tomita, M.; Doki, S.; Okuma, S. An extended electromotive force model for sensorless control of interior permanent-magnet synchronous motors. *IEEE Trans. Ind. Electron.* **2003**, *50*, 288–295.

22 Morimoto, S.; Kawamoto, K.; Sanada, M.; Takeda, Y. Sensorless control strategy for salient-pole PMSM based on extended EMF in rotating reference frame. *IEEE Trans. Ind. Appl.* **2002**, *38*, 1054–1061.

23 Ichikawa, S.; Tomita, M.; Doki, S.; Okuma, S. Sensorless control of synchronous reluctance motors based on extended EMF models considering magnetic saturation with online parameter identification. *IEEE Trans. Ind. Appl.* **2006**, *42*, 1264–1274.

24 Hasegawa, M.; Yoshioka, S.; Matsui, K. Position sensorless control of interior permanent magnet synchronous motors using unknown input observer for high-speed drives. *IEEE Trans. Ind. Appl.* **2009**, *45*, 938–946.

25 Ribeiro, L.A.d.S.; Harke, M.C.; Lorenz, R.D. Dynamic properties of back-EMF based sensorless drives. In Proceedings of the 41st IAS Annual Meeting on Industry Applications Conference, Tampa, FL, USA, 8–12 October 2006; pp. 2026–2033.

26 Foo, G.; Rahman, M.F. Sensorless direct torque and flux-controlled IPM synchronous motor drive at very low speed without signal injection. *IEEE Trans. Ind. Electron.* **2010**, *57*, 395–403.

27 Vogelsberger, M.A.; Grubic, S.; Habetler, T.G.; Wolbank, T.M. Using PWM-induced transient excitation and advanced signal processing for zero-speed sensorless control of ac machines. *IEEE Trans. Ind. Electron.* **2010**, *57*, 365–374.

28 Morimoto, S.; Sanada, M.; Takeda, Y. Mechanical sensorless drives of IPMSM with online parameter identification.*IEEE Trans. Ind. Appl.* **2006**, *42*, 1241–1248.

29 Mobarakeh, B.N.; Tabar, F.M.; Sargos, F.M. Back EMF estimation-based sensorless control of PMSM: Robustness with respect to measurement errors and inverter irregularities. *IEEE Trans. Ind. Appl.* **2007**, *43*, 485–494.

30 Inoue, Y.; Yamada, K.; Morimoto, S.; Sanada, M. Effectiveness of voltage error compensation and parameter identification for model-based sensorless control of PMSM. *IEEE Trans. Ind. Appl.* **2009**, *45*, 213–221.

31 Genduso, F.; Miceli, R.; Rando, C.; Galluzzo, G.R. Back EMF sensorless-control algorithm for high-dynamic performance PMSM. *IEEE Trans. Ind. Electron.* **2010**, *57*, 2092–2100.

32 Raute, R.; Caruana, C.; Staines, C.S.; Cilia, J.; Sumner, M.; Asher, G.M. Analysis and compensation of inverter nonlinearity effect on a sensorless PMSM drive at very low and zero speed operation. *IEEE Trans. Ind. Electron.* **2010**,*57*, 4065–4074.

33 Kim, H.; Lorenz, R.D. Synchronous frame PI current regulators in a virtually translated system. In Proceedings of the 39th IAS Annual Meeting on Industry Applications Conference, Seattle, WA, USA, 3–7 October 2004; pp. 856–863.

34 Kim, H.; Lorenz, R.D. A virtual translation technique to improve current regulator for salient-pole AC machines. In Proceedings of the 2004 IEEE 35th Annual Power Electronics Specialists Conference, Aachen, Germany, 20–25 June 2004; pp. 487–493.

35 Yang, S.M.; Su, P.D. Active damping control of hybrid stepping motor. In Proceedings of the 2001 4th IEEE International Conference on Power Electronics and Drive Systems, Denpasar, Indonesia, 22–25 October 2001; pp. 749–754.

36 Yim, J.S.; Sul, S.K.; Bae, B.H.; Patel, N.R.; Hiti, S. Modified current control schemes for high-performance permanent-magnet AC drives with low sampling to operating frequency ratio. *IEEE Trans. Ind. Appl.* **2009**, *45*, 763–771.

37 Sul, S.K. *Control of Electric Machine Drive System*, 1st ed.; Brain Korea: Seoul, Korea, 2002.

CHAPTER 3

Characteristics of and Control over Resonance in the Electromotive Force of Electromagnetic Induction

Sang Don Bu, Jin Kyu Han, Jin Young Hyeon, Gi Gwan Kim

Department of Physics, Chonbuk National University, Jeonju, South Korea.

ABSTRACT

The principles of electromagnetic induction are applied in many devices and systems, including induction cookers, transformers and wireless energy transfer; however, few data are available on resonance in the electromotive force (EMF) of electromagnetic induction. We studied electromagnetic induction between two circular coils of wire: one is the source coil and the other is the pickup (or induction) coil. The measured EMF versus frequency graphs reveals the existence of a resonance/anti-resonance in the EMF of electromagnetic induction through free space. We found that it is possible to control the system's resonance and anti-resonance frequencies. In some devices, a desired resonance or antiresonance frequency is achieved by varying the size of the resonator. Here, by contrast, our experimental results show that the system's resonance and anti-resonance frequencies can be adjusted by varying the distance between the two coils or the number of turns of the induction coil.

KEYWORDS

Electromagnetic Induction; Electromotive Force; Resonator, Resonance/Anti-Resonance Frequency

1. INTRODUCTION

1.1. Resonance

Resonance occurs widely in nature and is exploited in many man-made devices. Electric resonance is used in many circuits [1-5]; for example, radio and TV sets use resonance circuits to tune in to stations. In these devices, many frequencies reach the circuit simultaneously through the antenna, but significant current flow is induced only by frequencies at or near the circuit's resonance frequency. By varying the inductance or capacitance, the device can be tuned to different stations. In physics, resonance is the tendency of a system to oscillate at a greater amplitude at the system's resonance frequencies than at others. At these frequencies, even small periodic driving forces can produce large amplitude vibrations because the system stores the vibrational energy. In this work, we studied electromagnetic induction between two circular coils of wire: one is the source coil and the other is the pickup (or induction) coil, and report the characteristics of and control over the resonance and anti-resonance in the electromotive force (EMF) of electromagnetic induction through free space.

1.2. How Does the Magnitude of Electromotive Force Behave as the Frequency Applied to the Source Coil Increases?

The experiment was performed based on Faraday's law:

$$\varepsilon = -\frac{d\Phi}{dt}. \tag{1}$$

where ε and Φ are the EMF induced in the induction coil and the magnetic flux passing through the induction coil, respectively [1,6]. The experimental setup consists of two circular coils of wire composed of an electrically conductive copper wire of cross-sectional radius 0.35 mm tightly wound into a series of loops of 5 - 320 turns, radius 7 cm and height 2 cm, as shown in **Figure 1**. One coil (the source coil) is connected to a signal generator (AFG3021B, Tektronix), and the other (the pickup or induction coil) is connected to an oscilloscope (DSO 5012A, Agilent Technologies). The signal generator supplies a sinusoidal voltage to the source coil, creating a sinusoidal magnetic field. The AC magnetic field propagates through free space and reaches the pickup coil. According to Faraday's law, an electric field is induced in any region of space in which a magnetic field varies over time. Thus, an EMF is induced in the pickup coil. If the magnetic flux passing through the pickup coil is $\Phi = A\sin\omega t$, then the induced EMF is $\varepsilon = A\omega\cos\omega t$. As ω increases, ε increases, and the magnitude of $\mathcal{E}$ is proportional to the rate at which the magnetic flux changes with time, so that faster changes give a stronger $\mathcal{E}$. Here our question is: how does the magnitude of $\mathcal{E}$ behave as ω increases, especially in the high frequency range of 10 kHz to a few MHz?

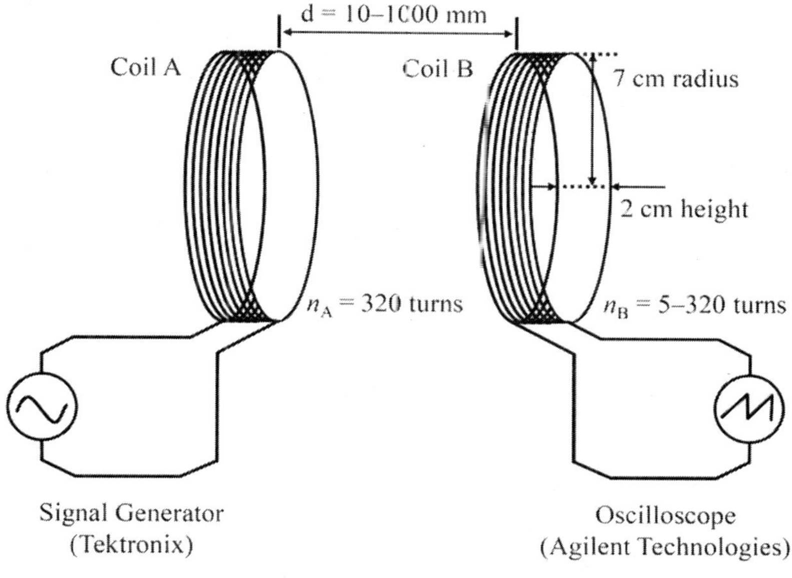

Figure 1. Schematic diagram of the experimental setup. The experimental setup consists of two coils. Coil A is the source coil, which transmits an electromagnetic wave through free space. Coil B is the pickup (or induction) coil, which detects an induced EMF. Coils A and B are aligned coaxially, such that the distance d between the two coils can be adjusted from 10 mm to 1000 mm. The coils are immobile during operation.

2. EASE OF USE ELECTROMAGNETIC INDUCTION PROPERTIES

2.1. Comparison of the Experimental and Predicted Electromotive Forces

Figures 2(a) and (b) show typical behaviors of 1) the root-mean-square value of ε $(\varepsilon_{\mathrm{RMS}})$ of the pickup coil and 2) the phase difference φ between the applied voltage (to the source coil) and the generated ε (in the pickup coil), respectively, as a function of the applied frequency f_Λ (to the source coil). We expected that, according to Faraday's law and the radiation resistance of the coil, $\varepsilon_{\mathrm{RMS}}$ would increase with increasing f_Λ and then finally attain a new equilibrium, as shown by the black dashed line in **Figure 2**(a). Such a response could be modeled using the Langevin function [7],

$$L(f_\Lambda) = \coth f_\Lambda - \frac{1}{f_\Lambda}. \tag{2}$$

However, the experimental data exhibited a very different behavior, as shown by the red solid circles in Figure 2(a). A resonance behavior was observed at high frequencies, and a relaxation behavior was observed at low frequencies. Resonance and anti-resonance peaks were clearly observed at 87 kHz and 285 kHz, respectively.

Figure 2(b) shows that at a frequency of 10 Hz, the ε of the pickup coil was 78° out-of-phase with respect to the wave applied to the source coil. As f_Λ increased, the phase difference decreased until the phases matched, where the phase-matching means that the phase difference is less than 5°. Meanwhile, the phase difference changed profoundly when the pickup coil became resonant. The ε of the pickup coil was 176° out-of-phase at frequencies between the resonance and anti-resonance peaks, that is, over a frequency range of 87 - 285 kHz.

The abrupt change in the phase difference may have been due to motions of the charges in the pickup coil. If the phase difference arose from the acceleration of conducting electrons induced by the electromagnetic radiation (generated from the source coil), the phase difference may be effectively independent of f_A at high frequencies. However, if the phase difference arose from the oscillations in the electric/magnetic dipoles or toroidal dipoles[1] [8-14] induced by the electromagnetic radiation, the phase difference would be expected to depend on f_A at high frequencies.

The radiation resistance in a coil with n turns composed of an electrically conductive copper wire may be modeled as follows. For a coil of $n = 320$, radius 7 cm and height 2 cm, as used here (see**Figure 1**), the electric dipole radiation term in the radiation resistance is smaller than the magnetic dipole radiation term, assuming that the distance from the coil is $r = c/f$, where c is the speed of light, for example, $r = 1000$ mm; the former is on the order of $(\omega/1000c)^2$, the latter is on the order of $(10\omega/c)^4$, where ω and c are the angular frequency and the speed of light, respectively. The resonance frequencies obtained in our experiments were much lower than the resonance frequencies of the valence electrons or electric/magnetic dipoles [6,7]. Therefore, in the particular geometry studied here, the abrupt changes in the phase difference may have predominantly arisen from oscillations in the toroidal dipoles [13,14] in the induction coil.

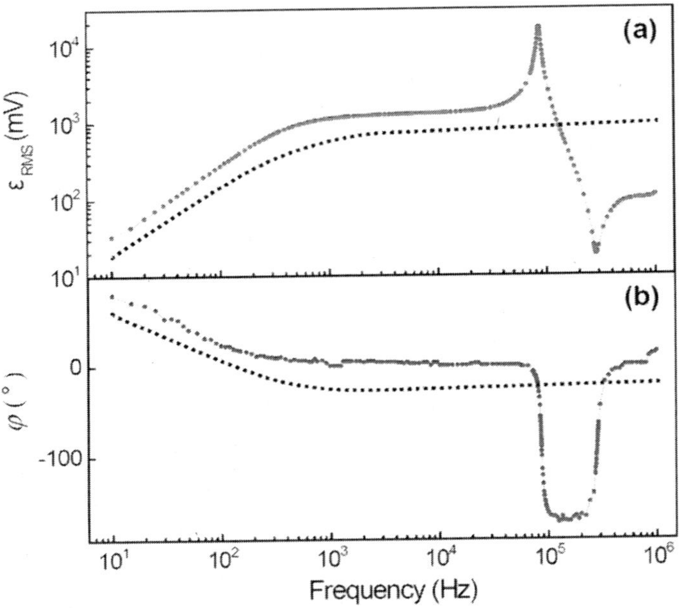

Figure 2. Comparison of the experimental and predicted values of ε_{RMS} and φ. The experimental measurements of ε_{RMS} and φ are shown in (a) and (b), respectively, indicated by red and blue solid circles. The amplitude of the signal applied to the source coil and the distance between the source and induction coils were then fixed at 14 V and 3 cm, respectively. The predicted values of ε_{RMS} and φ are shown in (a) and (b), respectively, with black dashed lines.

2.2. Adjusting of Resonance and Anti-Resonance Frequencies

To determine whether the resonance frequency f_R and anti-resonance frequency f_{AR} could be adjusted, we examined the influence of three experimental parameters on the f_R and f_{AR}. Figures 3(a)-(c) show graphs of ε_{RMS} versus f for various 1) voltages applied to the source coil (V_A = 2, 6, 10 and 14 V), 2) distances between the two coils (d = 10, 30, 100 and 1000 mm), and 3) numbers of turns of the pickup coil (n_B = 5, 50, 150, 250 and 320), respectively. Figures 4(a)-(c) show variations of f_R and f_{AR} for V_A, d and n_B, respectively, obtained from **Figure 3**. The resonance and anti-resonance peaks are clearly evident in most of the curves in **Figure 3**. f_R was not influenced by V_A or d, as shown by the black squares in Figures 4(a)-(b). The interval Δf_{R-AR} between f_R and f_{AR} did not change significantly with increasing V_A (**Figure 4**(a)), whereas Δf_{R-AR} decreased dramatically as d increased (**Figure 4**(b)). These results indicate that f_{AR} could be controlled by d independently of f_R. On the other hand, as n_B increased, f_R and f_{AR} decreased together, as shown in **Figure 4**(c).

f_R and f_{AR} each show peculiar behavior. The f_R and f_{AR} curves shown in **Figure 4**(b) indicate that f_R does not depend on d and f_{AR} could be described by an equation involving the logarithm of d. A plot of f_{AR} as a function of $\log d$ yielded a straight line, as indicated by the red line in the inset of **Figure 4**(b). Thus, f_{AR} can have the form "$m_1 \log d + c_1$", where $m_{i=1,2,3,4,5}$ and $c_{i=1,2,3,4,5}$ are the proportionality constants and the vertical axis intercepts (or meaningless constants), respectively. According to **Figure 4**(c), f_R can have the form "$m_2 n_B + c_2$", rather than "$m_3 \log n_B + c_3$", and f_{AR} can also have the form "$m_4 n_B + c_4$". The experimental results obtained for various combinations of V_A, d and n_B show that f_{AR} could be described by an equation of the form "$m_1 \log d + m_4 n_B + m_5 V_A + c_5$" with the condition $|m_1| \gg |m_4| \gg |m_5|$, and f_R could be described by an equation of the form "$m_2 n_B + c_2$" with the condition $|m_2| \cong |m_4|$. It is noted that under the conditions of various n_B turns, all the behaviors of f_R and f_{AR} with d were like the results shown in **Figure 4**(b), indicating that the cross term "dn_B" might be negligible.

We showed experimentally that pickup coils with different resonance/anti-resonance frequencies could be obtained by selecting the appropriate d and n_B for a given coil size. These results suggest the possibility of "a wireless power transfer station" that transmits power from a source coil to a large number of pickup coils through free space. If several pickup coils (with different resonance/anti-resonance frequencies) were positioned around a source coil (i.e., a short-range power networking system), power could be transferred from the source coil to the pickup coils by modulating the rate of change (over time) of the magnetic flux passing through each pickup coil. Wireless power transfer stations that are analogous to radio stations may be realized in the near future to permit everyone to use power anywhere without the need for wired power transmission.

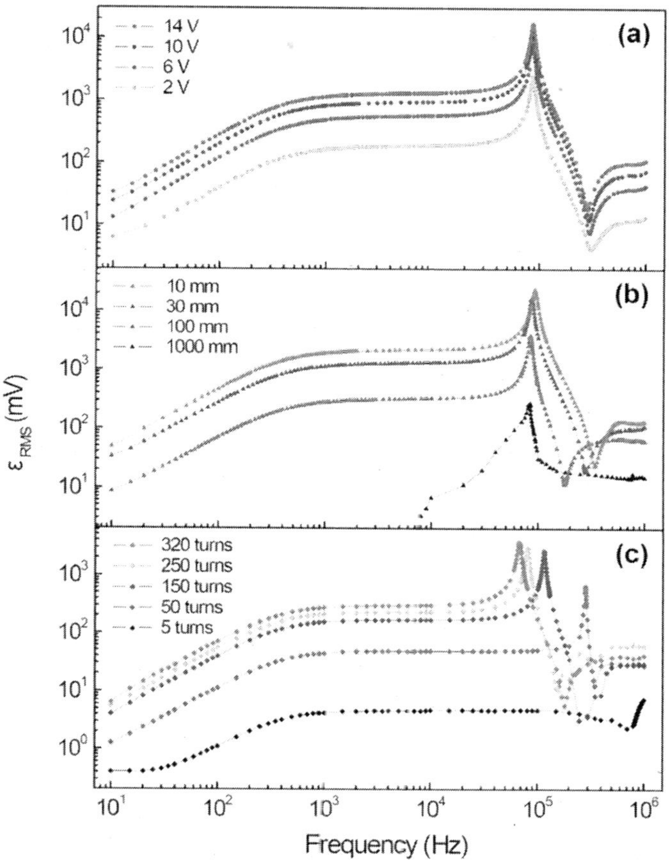

Figure 3. Plots of ε_{RMS} versus f_A. ε_{RMS} was measured as a function of the voltage applied to the source coil (V_A), the distance between the two coils (d) and the number of turns of the pickup coil (n_B), where the number of turns of the source coil (n_A) was fixed at 320. (a) Plots of ε_{RMS} versus $f_A$for V_A of 2, 6, 10 and 14 V, with d and n_B fixed at 30 mm and 320, respectively; (b) Plots of ε_{RMS} versus f_A for d of 10, 30, 100 and 1000 mm, with V_A and n_B fixed at 14 V and 320, respectively; (c) Plots of ε_{RMS} versus f_A for n_B of 5, 50, 150, 250 and 320, with V_A and d fixed at 14 V and 100 mm, respectively.

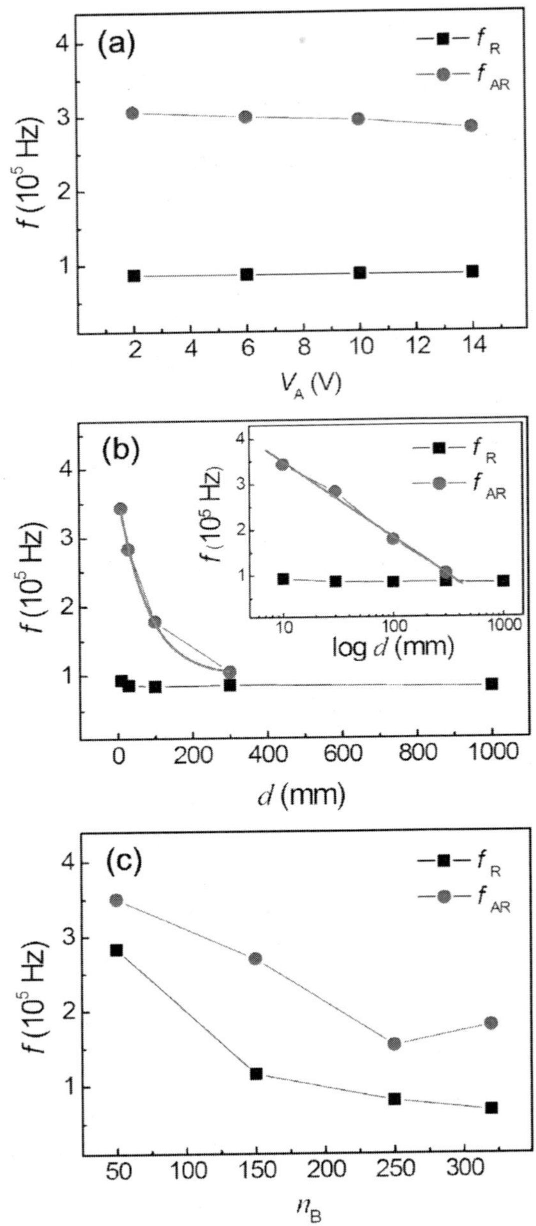

Figure 4. f_R and f_{AR} as a function of (a) V_A, (b) d and (c) n_B.

3. CONCLUSION

In summary, we have studied the electromagnetic induction between two circular coils of wire and showed clearly the existence of resonance/anti-resonance in EMF of electromagnetic induction through free space. We believe that our results might provide a competitive approach toward the development of high-efficiency systems in devices of induction cookers, electric power transformers, and wireless energy transfer.

4. ACKNOWLEDGEMENTS

This research was supported by Basic Science Research Program through the National Research Foundation of Korea (NRF) funded by the Ministry of Education, Science and Technology (grant number 2012R1A1A 2042743).

REFERENCES

1 D. C. Giancoli, "Physics: Principles with Applications," Prentice Hall, Upper Saddle River, 2005, pp. 584-608.

2 A. Kurs, A. Karalis, R. Moffatt, J. D. Joannopoulos, P. Fisher and M. Soljacic, "Wireless Power Transfer via Strongly Coupled Magnetic Resonances," Science, Vol. 317, No. 5834, 2007, pp. 83-86.

3 T. Imura, T. Uchida and Y. Hori, "Flexibility of Contactless Power Transfer using Magnetic Resonance Coupling to Air Gap and Misalignment for EV," World Electric Vehicle Journal, Vol. 3, 2009, pp. 1-10.

4 S. A. Hackworth, X. Liu, C. Li and M. Sun, "Wireless Solar Energy to Homes: A Magnetic Resonance Approach," International Journal of Innovations in Energy Systems and Power, Vol. 5, No. 1, 2010, pp. 40-44.

5 B. Wang, T. Nishino and K. H. Teo, "Wireless Power Transmission Efficiency Enhancement with Metamaterials," Proceedings of Wireless Information Technology and Systems of 2010 IEEE International Conference, Honolulu, 28 August-3 September 2010, pp. 1-4.

6 J. D. Jackson, "Classical Electrodynamics," Wiley, New York, 1999.

7 R. Coelho, "Physics of Dielectrics for the Engineer," Elsevier, New York, 1979, pp. 25-31, 62-73.

8 V. M. Dubovik, M. A. Martsenyuk and B. Saha, "Material Equations for Electromagnetism with Toroidal Polarizations," Physics Review E, Vol. 61, No. 6, 2000, pp. 7087-7097.

9 V. M. Dubovik, B. Saha and J. L. Rubin, "Lorentz Transformation of Toroid Polarization," Ferroelectrics Letters Section, Vol. 27, No. 1-2, 2000, pp. 1-6.

10 G. N. Afanasiev, "Simplest Sources of Electromagnetic Fields as a Tool for Testing the Reciprocity-Like Theorems," Journal of Physics D: Applied Physics, Vol. 34, No. 4, 2001, pp. 539-559.

11 V. M. Dubovik and V. V. Tugushev, "Toroid Moments in Electrodynamics and Solid-State Physics," Physics Reports, Vol. 187, No. 4, 1990, pp. 145-202.

12 K. Marinov, A. D. Boardman, V. A. Fedotov and N. Zheludev, "Toroidal Metamaterial," New Journal of Physics, Vol. 9, No. 9, 2007, pp. 324-202.

13 A. A. Gorbatsevich and Yu. V. Kopaev, "Toroidal Order in Crystals," Ferroelectrics, Vol. 161, No. 1, 1994, pp. 321-334.

14 H. Schmid, "Toroidal Moments in Spin-Ordered Crystals," Proceedings of One Day International Research Workshop on Super-Toroidal Electrodynamics, University of Southampton, Southampton, 2004, pp. 108-177.

NOTES

[1]Toroidal moments were first considered by Zel'dovich in 1958. They are fundamental electromagnetic excitations that cannot be represented in terms of the standard multipole expansion. The toroidal dipole moment represents the lowest moment of the third independent electromagnetic multipole family, after the electric and magnetic moments. The properties of materials that possess toroidal moments and the classification of their interaction with external electromagnetic fields are topics of growing interest [8-14].

CHAPTER 4

Analysis and Design of a Transverse-Flux Dual Rotor Machine for Power-Split Hybrid Electric Vehicle Applications

Ping Zheng *, Quanbin Zhao, Jingang Bai, Bin Yu, Zhiyi Song and Jing Shang

Department of Electrical Engineering, Harbin Institute of Technology, Harbin 150080, China

ABSTRACT

A novel compound-structure transverse-flux permanent-magnet synchronous machine (CS-TFPMSM) is proposed in this paper, which is used in hybrid electric vehicles (HEVs) to fulfill the power-split function. The key component of the CS-TFPMSM is a brushless transverse-flux dual rotor machine (TFDRM). The TFDRM originates from the transverse-flux machines, and is capable of speed adjustment between the transverse-flux rotor and the permanent-magnet rotor without using any brushes. The structure and principle of the TFDRM are described. The torque equations of the TFDRM are deduced, which are different from those of traditional machines. Based on the investigation, the TFDRM tends to have high leakage and a poor power factor. The method to obtain high power factor is discussed. The back electromotive force (BEMF) and torque of the TFDRM are simulated with the variation of parameters, such as pole-pair number, width of the permanent magnets (PMs), and so on. A prototype of a 10kW TFDRM is designed

KEYWORDS

hybrid electric vehicle; transverse-flux; dual rotor machine; brushless; speed adjustment

1. INTRODUCTION

Hybrid electric vehicles (HEVs) have received much attention in view of their low emissions and high fuel economy. The first commercially available HEV, the Toyota Prius, was marketed in 1997, and utilizes a planetary gear and two electrical machines to fulfill the power-split function. The planetary-gear power-split system has proven to be a successful configuration, but the planetary gear, as a mechanical device, inevitably causes problems of transmission loss, gear noise, and it needs regular lubrication. Therefore, compound-structure machines have been explored to replace the planetary-gear power-split system [1,2].

Among them, the compound-structure permanent-magnet synchronous machine (CS-PMSM) has been widely investigated, demonstrating the ability to be an alternative to the planetary gear [3]. The CS-PMSM employs two electrical machines to take full control of the output torque and speed of the internal combustion engine (ICE). By adopting a suitable control strategy, the CS-PMSM can make the ICE operate at optimum efficiency almost independent of road conditions, thus maximizing the system efficiency [4]. However, the CS-PMSM has brushes and slips, leading to the disadvantages of friction losses, high maintenance, need for cooling and so on. As a consequence, new brushless compound-structure machines are being developed to avoid the shortcomings of the CS-PMSM.

In this paper, a brushless compound-structure transverse-flux permanent-magnet synchronous machine (CS-TFPMSM) is proposed. The key component of the CS-TFPMSM is a brushless transverse-flux dual rotor machine (TFDRM) which originates from the principle of transverse-flux machines. The transverse-flux machine refers to the electrical machine that has a magnetic circuit which is perpendicular to the direction of current flow and rotor motion [5].

Different from traditional machines, because the space occupied by the electrical and magnetic paths have no competition to each other, the transverse-flux machine can offer high power density and thus make the system smaller [6], which is suitable for electric vehicle applications [7]. The transverse-flux machine has some disadvantages though, such as a complicated structure with 3D magnetic circuit, high flux leakage, low power factor, difficult manufacture, and so on [8]. To get rid of these problems, many transverse-flux machines topologies have been investigated. According to the different arrangements of phases, windings, and permanent magnets (PMs), the transverse-flux machine can be divided into several types: in-plane phases, axial-arranged phases;

single-sided windings, double-sided windings; surface-mounted PMs, flux-concentration PMs; and so on [9]. The flux-concentration mode has the particular distinguishing characteristic that the flux density from the PMs can be concentrated and strengthened, which is favorable for the torque density [10,11]. Therefore, the TFDRM design is based on the flux-concentration transverse-flux machine.

2. BRIEF INTRODUCTION OF THE CS-TFPMSM

First of all, we introduce the schematic diagram of the CS-TFPMSM system. As shown in Figure 1, this system is mainly formed by the CS-TFPMSM, two inverters, and the battery. It is placed between the ICE and wheels to realize the power-split function. The CS-TFPMSM has two axially arranged machines: a brushless transverse-flux dual rotor machine (TFDRM) and a torque-adjusting machine (TAM). The inner rotor (permanent-magnet rotor, PMR) of the TFDRM is directly connected to the ICE; and the outer rotor (transverse-flux rotor, TFR) of the TFDRM is connected to the rotor of the TAM and the wheels. The TFDRM is the part that possesses a transverse-flux structure, and the TAM is just a traditional PMSM. Like the transverse-flux machines, the TFDRM adopts the flux-concentration (FC) mode to arrange the PMs [12].

During operation, the TFDRM changes the speed of the output shaft of the TFR relative to the ICE, and transfers the torque from the ICE to the TAM. The TAM supplies the difference of the torques between the ICE and the wheels. In this way, the CS-TFPMSM fulfills the control on both the speed and torque, and allows the ICE to operate at optimum efficiency points almost independent of road conditions, hereby decreasing emissions and improving fuel economy. The energy flow in the brushless CS-TFPMSM system is illustrated in Figure 2.

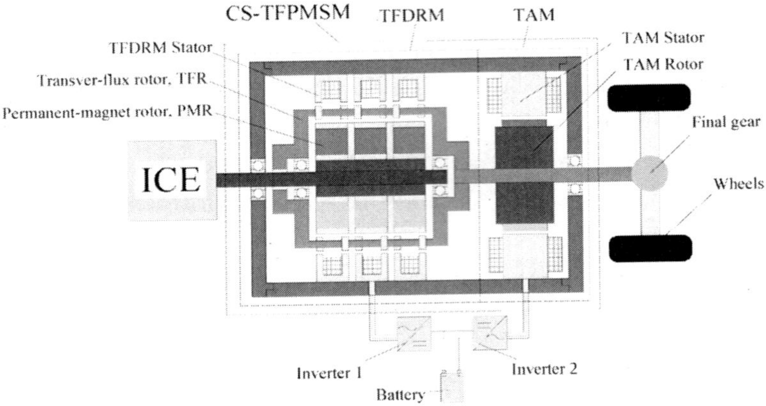

Figure 1. The CS-TFPMSM system.

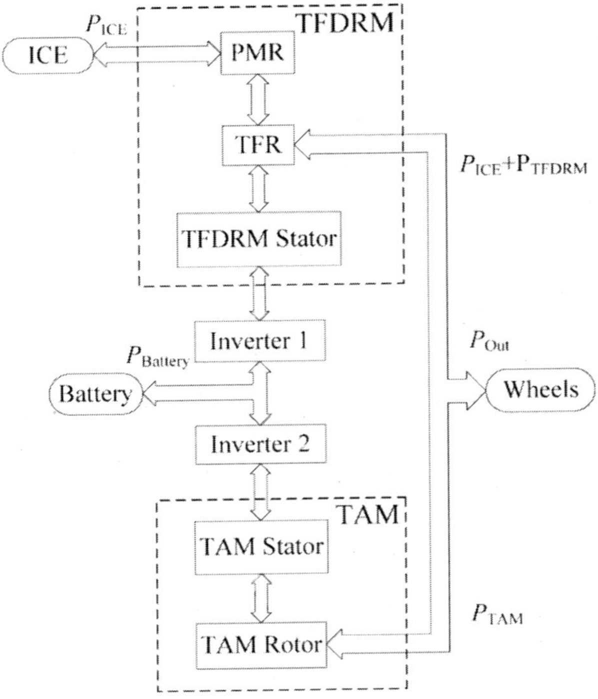

Figure 2. Energy flow in the CS-TFPMSM system.

There are six typical operation states of this system, as shown in Figure 3. Figure 3a illustrates the pure electric mode, in which the vehicle wheels are simply driven by the TAM. Figure 3b shows the ICE starter mode, in which only the TFDRM works. Figure 3c shows the braking mode, in which the TAM operates as a generator to recover the brake energy. Figure 3d shows a complex mode, hill-climbing mode. Because the requirement of wheels for torque is higher than the torque input from the ICE, the rest of the torque is provided by the TAM. Figure 3e shows another complex mode, the continuously variable transmission (CVT) mode, in which the TFDRM generates electric power to supply the TAM, and the speed and torque of the ICE are adjusted by both TFDRM and TAM to meet the requirement of the wheels. Figure 3f also shows a complex mode, acceleration mode, in which the battery provides the power through the CS-TFPMSM to satisfy the requirement of wheels for power.

To verify the power conversion, the relationship is derived as follows. In the TFDRM, the magnetic field generated by stator windings is modulated by the TFR and imposed on the PMR. Following the principle of the transverse-flux machines, there is a relative motion between the TFR and the PMR. The relationship between the speed of the TFR and the PMR is shown in Equation (1):

$$\Omega_{TFR} = \frac{\omega_{TAM}}{p_{TAM}} = \Omega_{PMR} + \frac{\omega_{TFDRM}}{p_{TFDRM}} \tag{1}$$

where Ω_{TFR} is the mechanical speed of the TFR (rad/s); Ω_{PMR} is the mechanical speed of the PMR (rad/s); ω_{TFDRM} is the angular frequency of the winding current in the TFDRM (rad/s); p_{TFDRM} is the pole-pair number of the TFDRM; ω_{TAM} is the angular frequency of the winding current in TAM (rad/s) and p_{TAM} is the pole-pair number of the TAM.

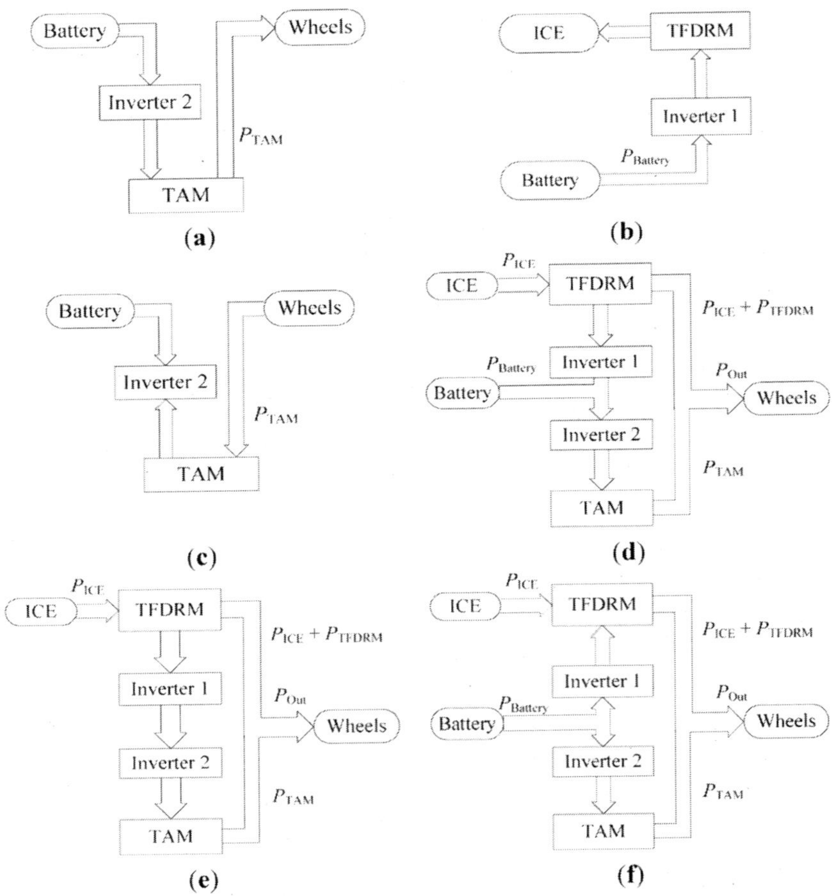

Figure 3. Typical operation states of the power-split hybrid drive system: (**a**) Pure electric mode; (**b**) ICE starter mode; (**c**) braking mode; (**d**) hill climbing mode; (**e**) CVT mode; (**f**) acceleration mode.

On the other hand, as for the output torque of the shaft, it comprises the torque of the TFDRM and the torque of the TAM:

$$T_{out} = T_{TAM} + T_{TFDRM} \tag{2}$$

where T_{TFDRM} is the torque of the TFDRM and T_{TAM} is the torque of the TAM which can be positive or negative depending on the difference between the demanded output torque and the torque generated by the ICE.

The output mechanical power is shown in Equation (3):

$$\begin{aligned} P_{Out} &= P_{ICE} + P_{Battery} \\ &= P_{ICE} + P_{TFDRM} + P_{TAM} \\ &= T_{TFDRM} \cdot \Omega_{PMR} + T_{TFDRM} \cdot \frac{\omega_{TFDRM}}{p_{TFDRM}} + T_{TAM} \cdot \left(\Omega_{PMR} + \frac{\omega_{TFDRM}}{p_{TFDRM}} \right) \\ &= \left(T_{TFDRM} + T_{TAM} \right) \cdot \left(\Omega_{PMR} + \frac{\omega_{TFDRM}}{p_{TFDRM}} \right) \\ &= T_{out} \cdot \Omega_{TFR} \end{aligned} \tag{3}$$

where P_{TFDRM} is the electromagnetic power of the TFDRM; P_{TAM} is the electromagnetic power of the TAM; $P_{Battery}$ is the input power from the battery; and P_{ICE} is the input mechanical power from the ICE.

In all, the TFDRM and the TAM of the brushless CS-TFPMSM are employed to adjust the differences of speed and torque between the ICE and the wheels. Without any brushes or planetary gears, both maintenance cost and mechanical cogging noise of the HEV system are greatly reduced. This advantage makes the CS-TFPMSM favorable for HEV applications.

3. STRUCTURE AND PRINCIPLE OF THE TFDRM

3.1. The Structure of the TFDRM

The TFDRM is the key component of the CS-TFPMSM, and originates from the transverse-flux machines. Like the axial three-phase transverse-flux machines, the TFDRM is actually made of three separate single phases. The three single phases of the TFDRM are axially stacked with a small gap between one another to decouple them magnetically and avoid flux leakage between phases. The electrical angle between two phases is 120°.

Figure 4 shows one phase of the 6-pole-pair TFDRM. Every phase consists of three parts radically arranged (from outside to inside): the stator, the TFR, and the PMR. There is an outer air gap between the stator and the TFR, and an inner air gap between the TFR and the PMR, too. Both the TFR and the PMR rotate peripherally. The stator core has a U-shaped cross section which contains a yoke and two endings. In the stator of every phase, there is a simple concentric toroidal armature winding, which is surrounded by the stator core.

Every phase has respectively corresponding transverse-flux teeth (TFT). The TFT of these three single phases are stacked in the axial direction, and are circumferentially shifted in space by $120/p_{\mathrm{TFDRM}}$ mechanical degrees (120 electrical degrees) to form a balanced three-phase topology. In one phase, the number of the TFT equals to the pole number. The TFT of one phase are equally distributed into two rows and fixed in a holder. To fit the arrangement of the permanent magnets in the FC mode, one row of the TFT should be circumferentially shifted by one pole pitch.

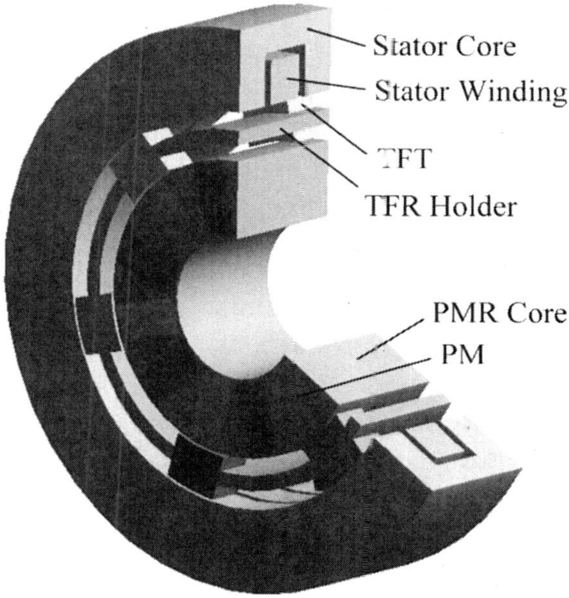

Figure 4. One phase of the TFDRM.

For every phase, there are also corresponding PMs in the PMR. The PMR of every phase is the same and located at an aligned position. In one phase, the number of the PMs also equals the pole number. Employing the FC mode, the magnets are circumferentially magnetized in an alternating way. In this setup, the steel parts between the successive opposite magnets of different polarity in the circumferential direction become the rotor poles. The FC mode utilizes all the magnets at any one time and this allows the flux density in the working air gap to increase significantly.

3.2. The Magnetic Circuit of the TFDRM

In every phase, the TFDRM has multiple flux loops magnetically coupled with the winding. Every flux loop is an integral element, going through the stator core, the TFT, the PMs and the PMR core. The magnetic circuit turns out to be quite a complex 3D path. Figure 5 illustrates the magnetic flux path for a two-pole section of the TFDRM in detail. The arrows represent the flux directions. For clarity, the other loops, which go through other magnets and couple with the winding, are not illustrated in the figure.

From the analysis of the flux path, we can find that the total number of flux loops is proportional to the pole number. This number can be determined without competing with the axial length of the stator slot, because the space occupied by the flux-carrying core irons and the armature windings are independent of other. With the relative motion between the TFR and the PMR, the flux-change rate is proportional to the product of the pole number and the relative speed between rotors.

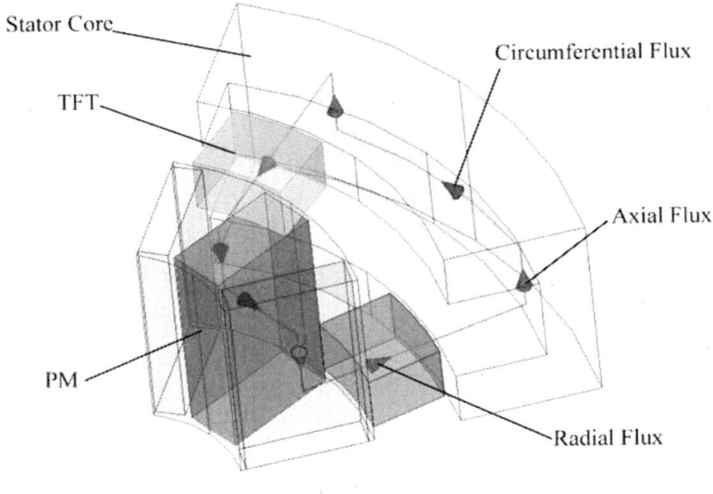

Figure 5. Magnetic flux path for a two-pole section.

Under load, every phase of the stator windings of TFDRM actually excites an independent one-pole-pair MMF, which is just an alternating wave rather than a common rotating field. Through the outer air gap, the one-pole-pair MMF is transferred into the corresponding TFT and transformed into a multi-pole-pair magnetic field. The interaction between the multi-pole-pair field in the inner air gap and the poles on the PMR will generate a torque. The torque of three phases is combined into a continuous effective torque leading to the difference between speeds of the TFR and the PMR.

From the former analysis on the magnetic circuit, we can see that the TFDRM actually possesses a magnetic circuit similar to the transverse-flux machines; and the brushes are eliminated through the shift of the structure of the transverse-flux machines.

Considered from another aspect, the TFDRM is developed from the transverse-flux machine in this way: the teeth of a transverse-flux machine are cut down and fixed in a holder to form the TFR; based on the remainder, a new stator is formed; and the PMR remains unchanged. Separation of the teeth from the stator of a transverse-flux machine produces a brushless dual rotor motor, with the principle of the transverse-flux machine unchanged.

In the aspect of manufacturing, the TFDRM also possesses some advantages: (1) the stator windings have a simple shape to manufacture and (2) a high slot fill factor can be achieved, so the space in the TFDRM can be used effectively.

4. CHARACTERISTICS OF THE TFDRM

4.1. Torque Equations of the TFDRM

To dimension the size of the TFDRM, the electric and magnetic loads are first determined in an analytical way. Although the concept of electric load remains available for the circular armature windings of the TFDRM, the general definition of electric load cannot properly reflect the real status that the windings are embedded in the inner surface of the stator of the TFDRM. So the electric load of the TFDRMs is redefined as the total ampere-turns per axial armature length:

$$A = \frac{NI}{l_{ef}} \tag{4}$$

where l_{ef} is the effective axial length of a single phase of the TFDRM; N is the number of the stator-winding turns; I is its current value.

The equation of the magnetic flux should be written as:

$$\Phi = \frac{\Phi_1}{\sigma_1 \sigma_2} = \frac{p B_{\delta 2} \alpha_p' \tau l_{ef}}{\sigma_1 \sigma_2} \tag{5}$$

$$\tau = \frac{\pi D_1}{2p} \tag{6}$$

where p is the pole-pair number; D_1 is the inner diameter of the TFR; σ_1 is the leakage factor of the flux from the outer air gap into the stator core; σ_2 is the leakage factor of the flux from the inner air gap into the PMR; Φ_1 is the magnetic flux in the inner air gap,

α'_p is the effective pole arc coefficient; and $B_{\delta 2}$ is the maximum flux density in the inner air gap.

As a result, the power equation of the TFDRM is derived as follows:

$$P_e = mEI\cos\psi = \frac{\sqrt{2}\pi^2 mD_1 l_{ef}^2 pnK_\Phi \alpha_p' B_{\delta 2} A\cos\psi}{120\sigma_1\sigma_2} \tag{7}$$

and the new torque density equation of the TFDRM is as follows:

$$\rho_T = \frac{T}{V} = \frac{\dfrac{P_e}{2\pi\dfrac{n}{60}}}{\dfrac{1}{4}\pi mD_0^2 l_{ef}} = \frac{\sqrt{2}D_1 l_{ef} pK_\Phi \alpha_p' B_{\delta 2} A\cos\psi}{\sigma_1\sigma_2 D_0^2} \tag{8}$$

where m is the phase number; n is the synchronous speed; D_0 is the outer diameter of the stator; ψ is the leading phase angle of the armature current I relative to E and K_Φ is the waveform factor of the magnetic flux.

From the equations above, we can find that the torque density of the TFDRM is proportional to the product of flux density B, which is the same as in the traditional PMSMs. As a result, the structure which can enhance the flux density B is favorable, for example, the FC mode [13].

The torque density of the TFDRM is also proportional to the pole-pair number, which is different from the traditional PMSMs. The reason is that the increase of the pole-pair number of the TFDRM results in a corresponding increase in the electrical frequency. On the other hand, the magnetic flux remains unchanged when the number of pole pairs is simply increased. The electric load also remains unchanged, because the increase of the pole-pair number has no impact on the available area of the stator slot. Consequently, there is an increase in the torque density.

As a result, in the design process, increasing the pole-pair number is a direct way to obtain a high power density, but there is a limitation in that the increase of pole-pair number can also probably increase the magnetic flux leakage among the TFT.

As a dual rotor machine, the TFDRM processes one more leakage factor σ_2. The existence of the parameter σ_2 increases the magnetic resistance of the main magnetic circuit, resulting in more flux leakage and a lower power factor. As a result, adopting a short air gap is favorable, but the length of the air gap is still limited by the level of processing technique.

4.2. Power Factor Analysis of the TFDRM

Like transverse-flux machines, the TFDRM tends to have a low power factor at high load. The characteristics of the TFDRM can be described and analyzed using a phasor diagram, the same as the traditional synchronous machines [14]. When the TFDRM is only fed with a current in the q-axis, and the d-axis armature current $I_d = 0$, the relationship of electromagnetic parameters is shown in Figure 6. I_q is the q-axis armature current. E_0 and E_δ are respectively the back electromotive forces (BEMFs) at the conditions of no-load and load, and U is the voltage fed by the external circuit. R_1 is the armature resistance, and X_q is the q-axis reactance. φ is the power factor angle, and θ is the torque angle.

With the small voltage drop across the phase resistance neglected, the power factor is expressed as:

$$\cos\varphi = \cos\left[\arctan\left(\frac{X_q I_q}{E_0}\right)\right] \tag{9}$$

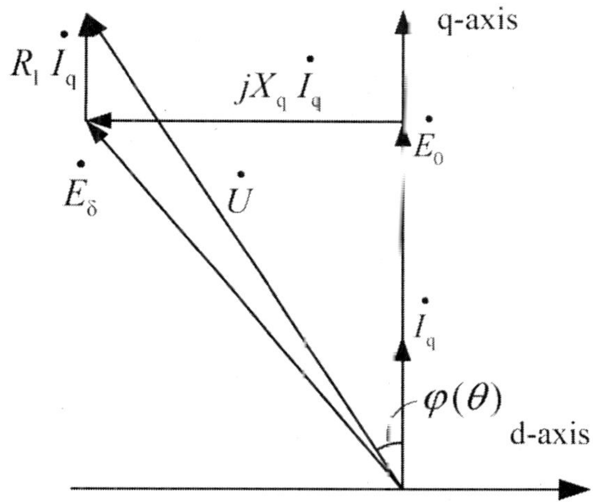

Figure 6. Phasor diagram when $I_d = 0$.

and if unsaturated:

$$\cos\varphi = \cos\left[\arctan\left(\frac{\Phi_i}{\Phi_m}\right)\right] \tag{10}$$

where Φ_m is the effective PM flux that links the stator winding and Φ_i is only the magnetic flux induced by the armature current.

From Equation (10), we know that increase the permanent magnetic flux Φ_m is favorable to improve the power factor. In the design stage, adopting some certain structures, for example, the FC mode, the magnetic flux density generated by permanent magnets that can be enhanced significantly. The FC mode is also beneficial to enhance the torque density, but the permanent magnets increase will lead to a high cost of materials, which makes this unattractive in practice.

Different from traditional machines and common transverse-flux machines, the TFDRM possesses two layers of air gaps. The two layers of air gaps could dramatically increase the reluctance of main magnetic circuit, and decrease magnetic flux Φ_m. Reduction of magnetic flux Φ_m will decrease the power factor. On the other hand, in the magnetic circuit of TFDRM, there is inherently a serious leakage flux due to the special transverse-flux configuration. The high leakage flux of the TFDRM also influences the power factor heavily. The enhancement of the reluctance of main magnetic circuit will increase the possibility of leakage flux. As a result, the two layers of air gaps make the problem of power factor much worse than common transverse-flux machines. Obviously, short length of outer/inner air gap is very important for increasing magnetic flux Φ_m and improving the power factor. But decreasing the length of outer/inner air gap is limited by the difficulty of manufacture. Finally, a compromise has to be done.

The equation can be further transformed into the following form:

$$\cos\varphi = \cos\left(\arctan\frac{X_q I_q}{E_0 - X_d I_d} - \psi\right) = \cos\left(\arctan\frac{X_q I\cos\psi}{E_0 - X_d I\sin\psi} - \psi\right) \quad (11)$$

where X_d and X_q are the d- and q-axis synchronous reactance. And $I_d = I \cdot \sin\psi$, $I_q = I \cdot \cos\psi$.

In the control of the TFDRM, there is also an effective way to improve the power factor by altering ψ, which is the leading phase angle of the armature current I relative to E_0. As shown in Figure 7, this adjustment can improve the power factor of the TFDRM, but it inevitably causes some sacrifice of the torque density. As for this method, we have to make a compromise between the power density and the torque density.

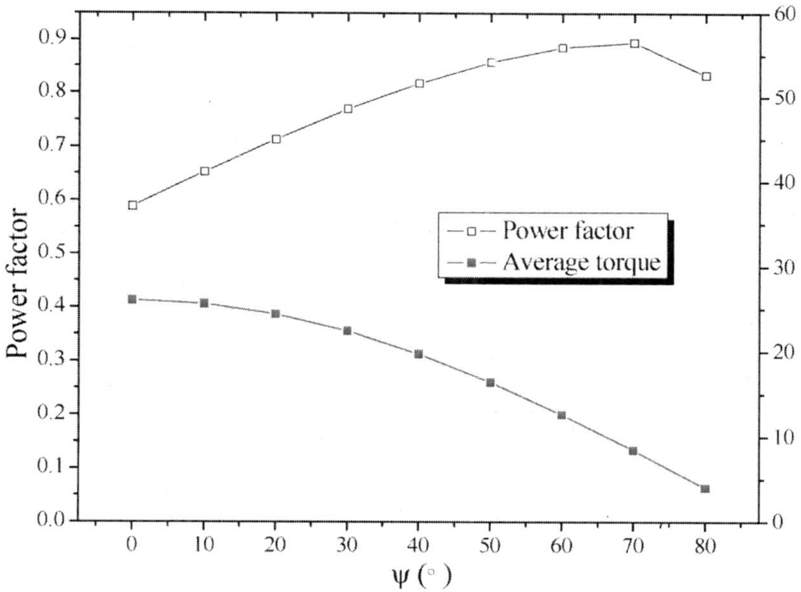

Figure 7. Power factor and average torque *versus* ψ.

5. MODELING AND SIMULATION OF THE TFDRM

The TFDRM with flux concentration has 3D flux paths which add complexity to the design and manufacturing of the machine [15,16]. To take the 3D effect into account and reflect the complex flux circuit and the flux leakage, 3D finite-element-method (FEM) simulation is employed to evaluate the performance of the TFDRM.

5.1. Determination of the 3D FEM Model

Because of the periodicity of the structure and the independence between the magnetic circuits of the three phases, only the structure of one pole pair in one phase needs to be modeled during the 3D FEM simulation. The simulation thus consumes less computation time and computer memory than the simulation of the total model. According to the structure, the 3D model of the TFDRM is set up and presented as follows (Figure 8).

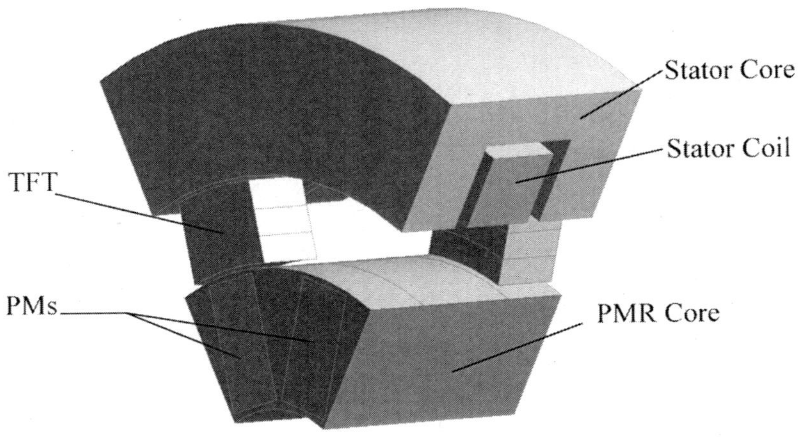

Figure 8. The 3D FEM model.

5.2. The Results of 3D FEM Simulation

With magneto transient 3D FEM simulation, we can directly obtain the back electric motion force (BEMF). Figure 9 shows the flux linkage and the fundamental BEMF at no-load.

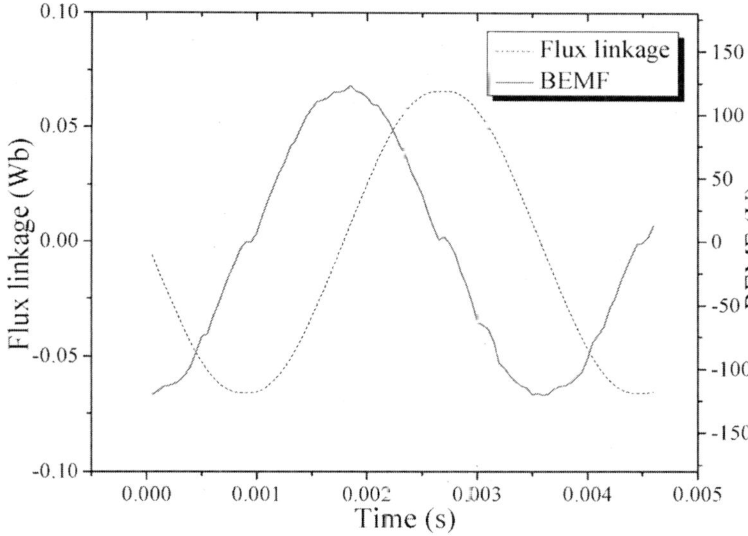

Figure 9. The flux-linkage and BEMF at no-load.

With the combination from the torques of the three single phases, we can obtain the torque curve and average torque [17].Figure 10 shows the torque curve at rated load.

6. OPTIMIZATION OF THE TFDRM

In the design of the TFDRM, high torque density is a major target to fulfill. Meanwhile, the power factor should be kept in a suitable range to avoid excessive control system capability. Considering these two aspects, the BEMF, average torque, and power factor *versus* main parameters are analyzed. The outline sizes are kept unchanged in the simulation.

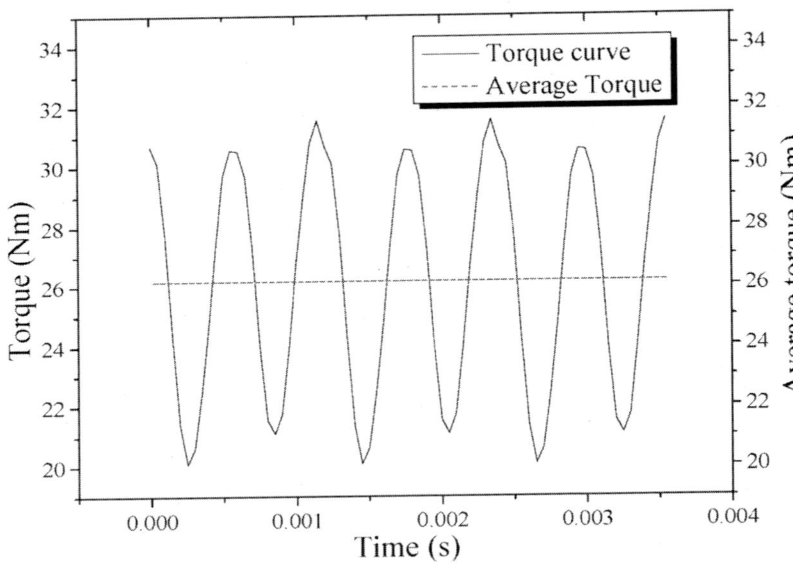

Figure 10. Torque curve from the 3D FEM simulation.

In this section, FEM simulations are used to compare the influence of different parameters on machine performance. When the influence of one parameter is analyzed, the other parameters are kept unchanged. The optimization sequence of parameters is based on their degree of influence on the performance.

The pole-pair number and the length of outer/inner air gap are optimized first. The pole-pair number influences the frequency and further the torque density of TFDRM, which is different from traditional machines. Besides, it will influence the sizes of TFT and PMs in the following design. The two layers of air gaps even make the TFDRM different from general transverse-flux machines. The length of outer/inner air gap has a significant impact on the reluctance of main magnetic circuit.

The axial length of stator-core ending is next to those two parameters. Increasing the axial length of stator-core ending will enhance the axial length of TFT and offer more area for the main magnetic circuit. However, when the axial length of one phase is unchanged, increasing the axial length of stator-core ending will decrease the axial length of stator slot and lead to smaller cross-sectional area of stator slot.

After that, the width of TFT and the thickness of PMs are determined. Increasing the width of TFT can offer more area for main magnetic flux; but the possibility of leakage flux among the TFT also increases. Larger thickness of PMs is favorable for the magnetic flux, but it will cut down the size of iron poles. As a result, these two parameters should be chosen in a suitable range according to the simulation results. Finally the width of PMs, which has a simple impact and can be determined more freely than other parameters, is optimized.

6.1. Pole-Pair Number

The pole-pair number is an important element in the optimization of the TFDRM. Generally speaking, the BEMF of the TFDRM increases with the pole-pair number, which is quite different from the traditional PMSMs. The simulation results are the same as the previous analysis, as shown in Figure 11.

To enhance the output power, we intend to increase the pole-pair number. However, this will probably lead to an increase of the leakage flux and a poor power factor. As shown in Figure 11 and Figure 12, when the pole-pair number is larger than 6, the torque increase no longer becomes significant, but the power factor drops quickly. The power factor reaches the highest value when the pole-pair number is 5. To obtain high torque density and maintain a suitable power factor, the pole-pair number should be in the 4~6 range. The core losses increase with the increase of pole-pair number.

In addition, high pole-pair number also increases the difficulty and cost of manufacture. In the TFDRM, the TFT are some small units. For the adopted structure with stator outside and rotor inside, the rotor surface space is limited. The size of the TFT will become even smaller because of a larger pole-pair number, which will make the manufacture and assembly difficult, so considering all these aspects, the pole-pair number is chosen to be 6.

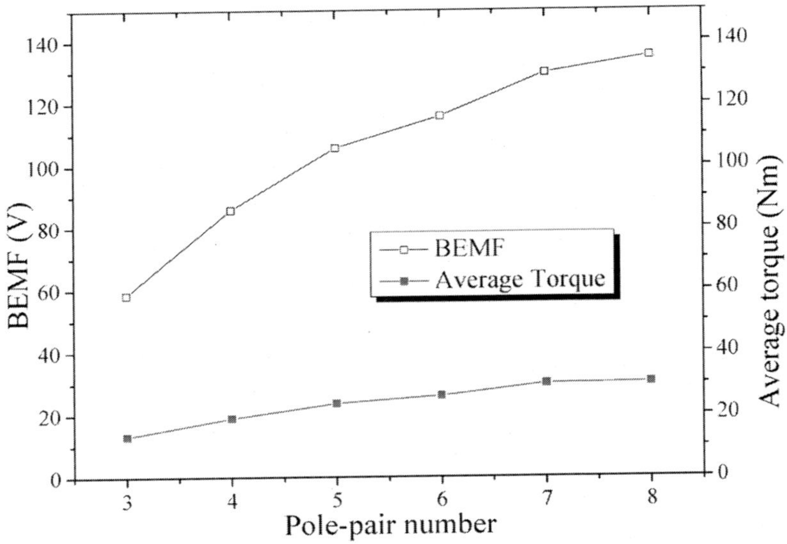

Figure 11. Variation of BEMF and average torque *versus* pole-pair number.

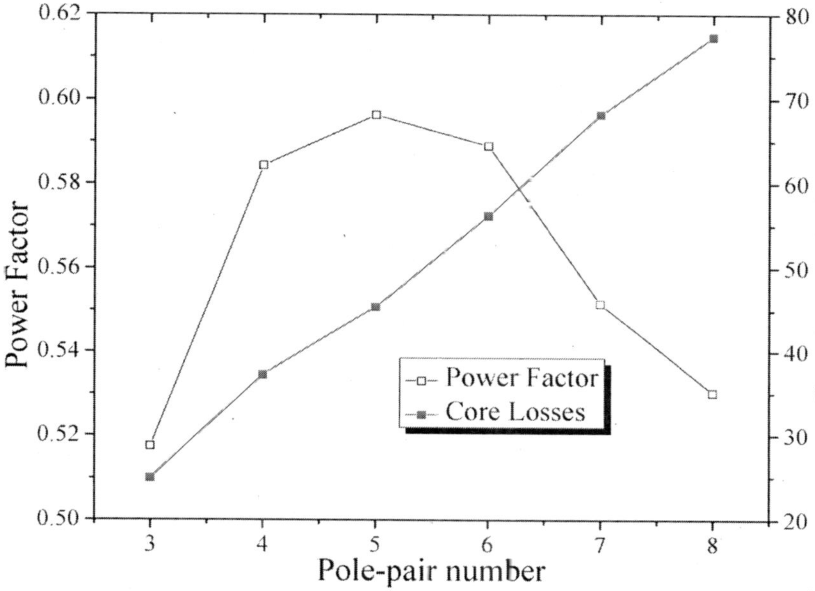

Figure 12. Variation of power factor and core losses *versus* pole-pair number.

6.2. Length of Outer/Inner Air Gap

During the optimization, the length of outer/inner air gap is one of the most important factors. In the TFDRM, two layers of air gaps make the reluctance of main magnetic circuit much larger. The influence of the length of outer/inner air gap on the BEMF, average torque and power factor is shown in Figure 13 and Figure 14.

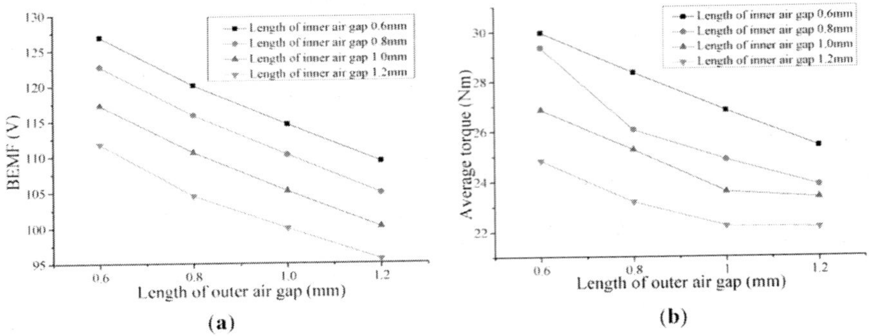

Figure 13. Variation of BEMF and average torque *versus* length of outer/inner air gap: (**a**) BEMF; (**b**) Average torque.

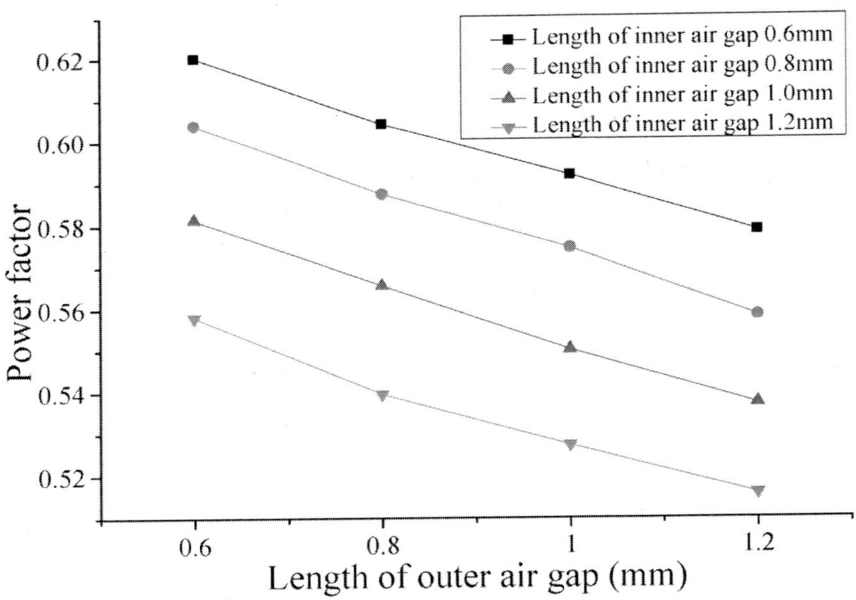

Figure 14. Variation of power factor *versus* length of outer/inner air gap.

The simulation results show that decreasing the length of outer/inner air gap is favorable for improving the performance of the TFDRM. Different from most other methods, using a shorter outer/inner air gap length can also increase torque density when the power factor is improved. The cost of materials does not increase either due to the saving of PMs. However decreasing the outer/inner air gap length will increase the difficulty and cost of manufacture and assembly, so as a result, the length of the outer/inner air gap is chosen to be 0.8 mm.

6.3. Axial Length of Stator-Core Ending

The stator core, which has a U-shaped cross section, is formed by two endings and a yoke. The increase of the axial length of the stator-core ending leads to a higher average torque. Figure 15 shows the effect of axial length of the stator-core ending on the BEMF and average torque. Increasing the axial length of stator-core ending is favorable for the power factor, because more area can be offered for the magnetic flux Φ_m, but increasing the axial length of stator-core endings leads to a corresponding decrease of the axial length of the stator slot, which increases flux leakage among stator endings and influences the power factor. As a consequence, these two aspects have to be balanced by adjusting the axial length of the stator-core ending. As shown in Figure 16, the 11~14 mm range is suitable for the axial length of stator-core ending. The length is chosen to be 13 mm.

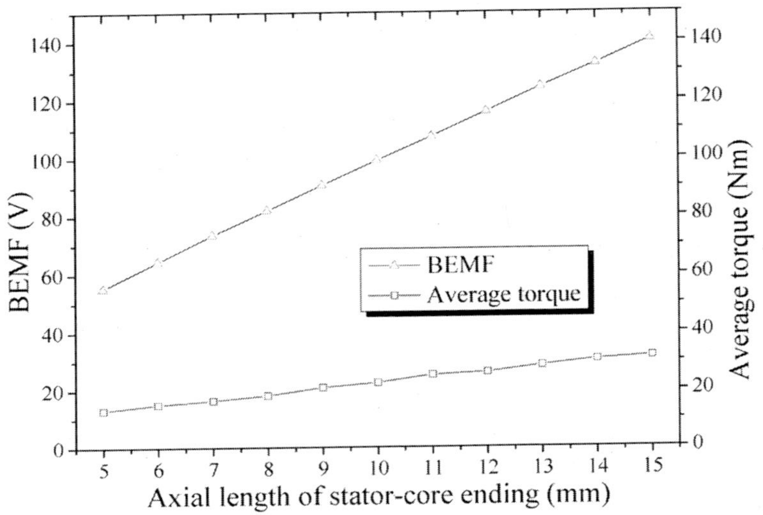

Figure 15. Variation of BEMF and average torque *versus* axial length of stator-core ending.

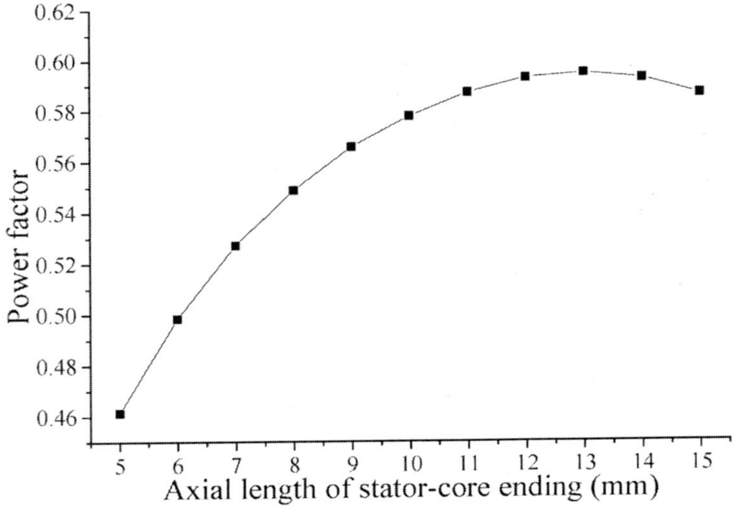

Figure 16. Variation of power factor *versus* axial length of stator-core ending.

6.4. Width of TFT

The size of the TFT has a critical impact on the characteristics of the TFDRM. The geometrical optimizations of the TFT are mainly aimed at enhancing the average torque and reducing the leakage flux between two adjacent teeth. The pole arc coefficient of the TFT internal surface, which is in contact with the inner air gap, has an effect on the BEMF and average torque as shown in Figure 17. Figure 17 shows that the average torque reaches higher levels when the width of TFT is within the 0.7–0.9 range.

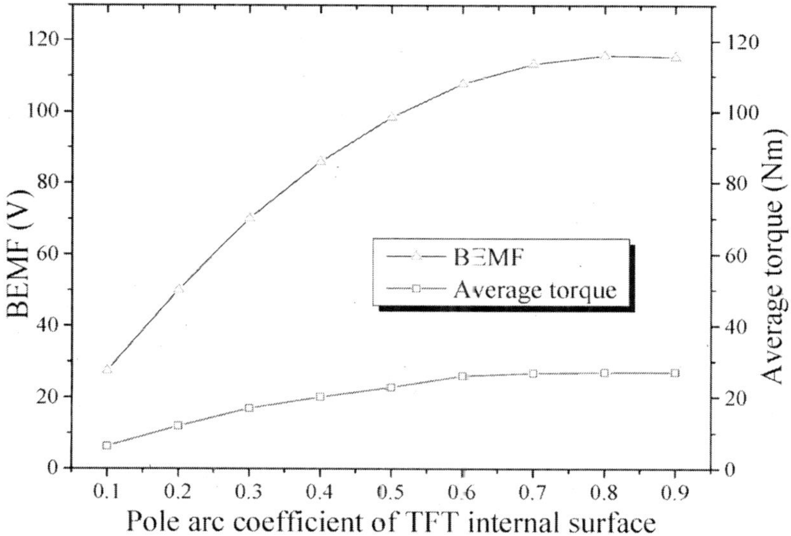

Figure 17. Variation of BEMF and average torque *versus* pole arc coefficient of the TFT internal surface.

With the pole arc coefficient increasing, the torque increases, but this increases the probability of flux leakage among the TFT. As a result, the torque increases slightly when the pole arc coefficient is beyond a certain value. Meanwhile, the leakage leads to a poor power factor. According to Figure 18, the width of TFT is chosen to be 0.8, which makes the power factor the highest.

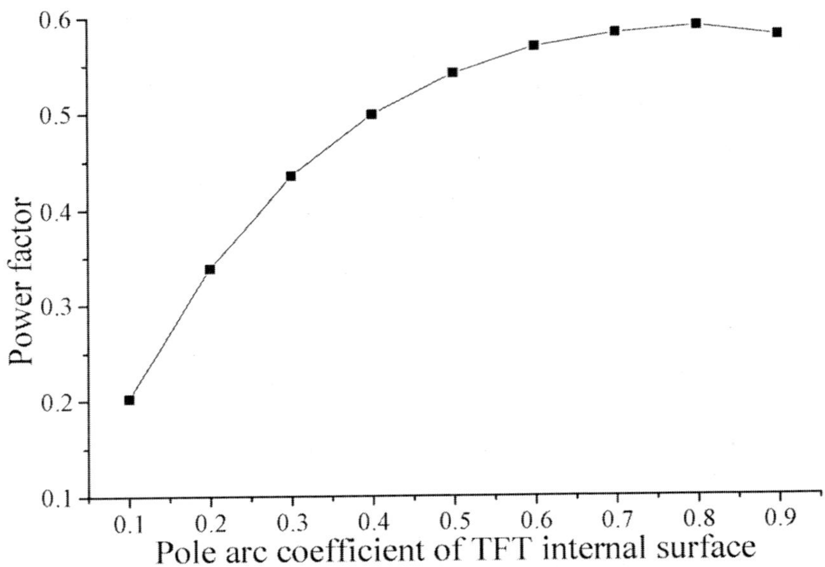

Figure 18. Variation of power factor *versus* pole arc coefficient of the TFT internal surface.

6.5. Thickness of PMs

In the TFDRM, the FC mode, which has distinguished characteristics, can also be adopted to concentrate and strengthen the flux density. Increasing the thickness of PMs can obtain higher BEMF, average torque and power factor, but the enhancement is slight when the thickness is bigger than 10 mm. The reason is that increasing the thickness will decrease the size of the iron pole. Consequently, when the thickness of PMs is too large, the magnetic flux concentration becomes worse; and the iron may face the problem of saturation. As shown in Figure 19 and Figure 20, the suitable value for the thickness of PMs is 10 mm.

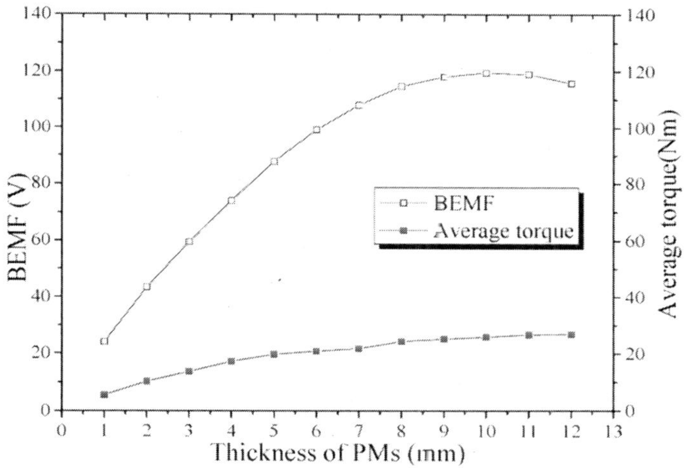

Figure 19. Variation of BEMF and average torque *versus* thickness of PMs.

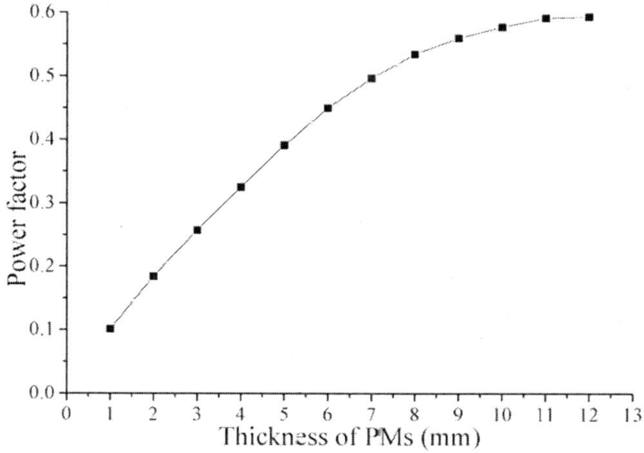

Figure 20. Variation of power factor *versus* thickness of PMs.

6.6. Width of PMs

Under the FC mode, to increase the width of PMs can increase the magnetic flux in the outer/inner air gap leading to a high torque. Figure 21 shows the variation of the BEMF and torque with the width of PMs. Increasing the width of PMs can also improve the power factor, as shown in Figure 22. The width of PMs has a simpler impact on the characteristics than other parameters. According to the simulation results, the width of PMs is chosen to be 27.8 mm.

Increasing the thickness and width of PMs can enhance torque density and power factor, but the cost of the TFDRM also increases, so the selection of the scheme is based on the balance of performance and cost.

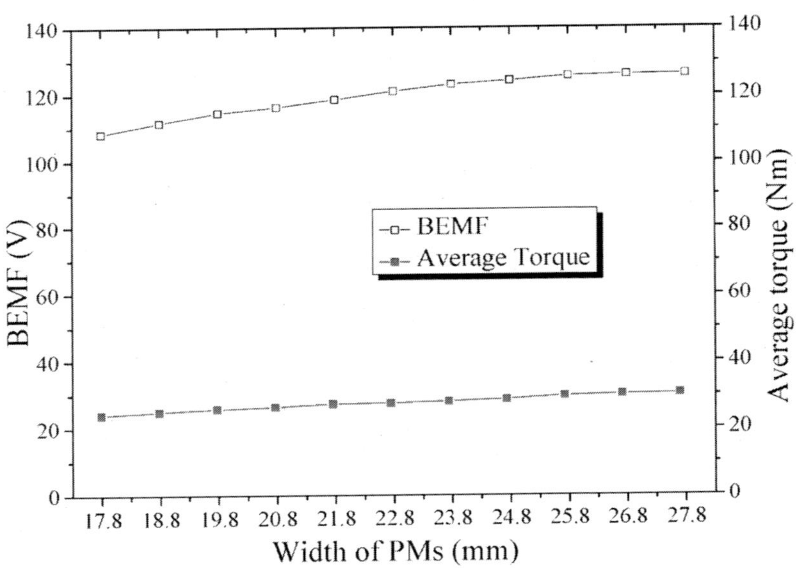

Figure 21. Variation of the BEMF and average torque *versus* width of PMs.

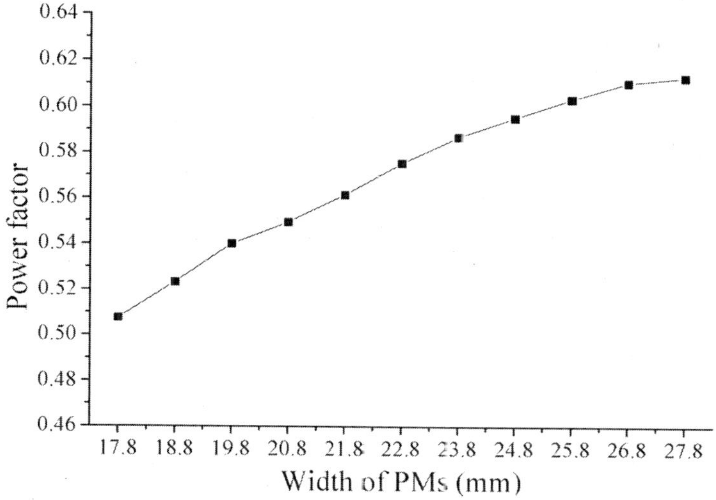

Figure 22. Variation of power factor *versus* width of PMs.

6.7. Comparison between Initial and Optimized Design

The parameters and performances of the initial and optimized schemes are compared in Table 1 and Table 2, respectively. As can be seen from Table 2, the BEMF, average torque, and torque density are all increased by using the optimized parameters. Especially, the power factor is enhanced significantly. Meanwhile, the THD of BEMF is also cut down.

6.8. Optimized Prototype of the TFDRM

A prototype of the TFDRM is designed and the parameters are shown in Table 3. The prototype will be manufactured, and the results of simulation will be verified by experimental measurements.

Table 1. Initial and Optimized Parameters of the TFDRM.

Parameters	Initial value	Optimized value
Pole-pair number	6	6
Stator outer diameter (mm)	203	189
Shaft diameter (mm)	48	38
Axial length of one phase (mm)	44	44
Outer air gap length (mm)	1.0	0.8
Inner air gap length (mm)	0.8	0.8
Axial length of stator-core ending (mm)	9	13
Axial length of stator slot (mm)	26	18
Pole arc coefficient of TFT internal surface	0.7	0.8
Axial length of TFT (mm)	9	13
Thickness of PMs (mm)	7	10
Width of PMs (mm)	20.8	27.8
Axial length of PMs (mm)	44	44

Table 2. Initial and Optimized Performances of the TFDRM.

Performances	Initial value	Optimized value
Rated power (kW)	10	10
Number of phase	3	3
Rated speed (rpm)	2800	2800
Rated current (A)	53.3	29.6
Rated power factor	0.28	0.61
Winding turns in series per phase	76	43
Amplitude of fundamental no-load BEMF (V)	70.2	126.2
THD of BEMF (%)	9.7	5.1
Average torque (Nm)	24.07	30.28
Torque density (Nm/m^3)	5641.1	8192.2

Table 3. Parameters of the TFDRM.

Parameters	Unit	Value
Rated power	kW	10
Number of phase	-	3
Rated speed	rpm	2800
Rated current	A	29.6
Number of poles	-	12
Rated efficiency	%	89.5
Rated power factor	-	0.61
Stator outer diameter	mm	189
Shaft diameter	mm	38
Axial length of TFDRM	mm	132
Outer air gap length	mm	0.8
Inner air gap length	mm	0.8
Winding turns in series per phase	-	43
Amplitude of fundamental no-load BEMF	V	126.2
THD of BEMF	%	5.1

7. CONCLUSIONS

In this paper, a new type CS-TFPMSM is proposed which is promising for power-split HEV applications. The CS-TFPMSM is capable of both speed adjustment and torque adjustment at the same time. As the key component of the CS-TFPMSM, the structure of the TFDRM is introduced, as well as the operation principles. The analysis shows that the TFDRM possesses the characteristics of a transverse-flux machine and the brushes can be eliminated. The torque equations of the TFDRM are deduced, which are different from those of traditional machines. The TFDRM has a complex 3D magnetic circuit and this may cause a low power factor. As a result, the way to obtain a high power factor is discussed: increasing the no-load BEMF is favorable for obtaining a higher power factor; making the leading phase angle of the armature current relative to no-load BEMF larger can efficiently improve the power factor with some sacrifice of torque. 3D FEM is used to simulate the influence of several parameters. The results indicate that increasing the axial length of stator-core ending or the width of the PMs can result in the enhancement of the BEMF and torque. Raising the pole-pair number and the pole arc coefficient of the TFT can also produce higher BEMF and torque; however, when these two parameters achieve a high level, the enhancement of the BEMF and torque becomes less significant. A 10 kW prototype is designed and optimized based on the analysis and simulation techniques.

ACKNOWLEDGMENT

This work was supported in part by National Natural Science Foundation of China under Projects 51077026, 51325701 and 51377030, and in part by the 863 Plan of China under Project 2011AA11A261.

CONFLICTS OF INTEREST

The authors declare no conflict of interest.

REFERENCES

1 Eriksson, S.; Sadarangani, C. A Four-Quadrant HEV Drive System. In Proceedings of the VTC 2002-Fall IEEE 56th Vehicular Technology Conference, Vancouver, BC, Canada, 24–28 September 2002; Volume 3, pp. 1510–514.

2 Nordlund, E.; Sadarangani, C. The Four-Quadrant Energy Transducer. In Proceedings of the 37th IAS Annual Meeting:

Industry Applications Conference, Pittsburgh, PA, USA, 13–18 October 2002; Volume 1, pp. 390–393.

3 Zheng, P.; Liu, R.; Wu, Q.; Tong, C.; Tang, Z. Compound-Structure Permanent-Magnet Synchronous Machine Used for HEVs. In Proceedings of the International Conference on Electrical Machines and Systems (ICEMS 2008), Wuhan, China, 17–20 October 2008; pp. 2916–2920.

4 Liu, R.; Zheng, P.; Zhao, H.; Sadarangani, C. Investigation of a Compound-Structure Permanent-Magnet Synchronous Machine Used for HEVs. In Proceedings of the IEEE Vehicle Power and Propulsion Conference (VPPC 2008), Harbin, China, 3–5 September 2008; pp. 1–5.

5 Arshad, W.M.; Backstron, T.; Sadarangani, C. Analytical Design and Analysis Procedure for a Transverse Flux Machine. In Proceedings of the IEEE International Electric Machines and Drives Conference (IEMDC 2001), Cambridge, MA, USA, 17–20 June 2001; pp. 115–121.

6 Huang, S.; Luo, J.; Lipo, T.A. Analysis and Evaluation of the Transverse Flux Circumferential Current Machine. In Proceedings of the 1997 IEEE Thirty-Second Industry Application Society Annual Meeting, New Orleans, LA, USA, 5–9 October 1997; pp. 378–384.

7 Schuttler, J.; Orlik, B. Analytical Model Describing the Operation Behavior of Transverse Flux Machines in Flat Magnet Configuration. In Proceedings of the 2007 European Conference on Power Electronics and Applications, Aalborg, Denmark, 2–5 September 2007; pp. 1–10.

8 Harris, M.R.; Pajooman, G.H.; Abu Sharkh, S.M. The Problem of Power Factor in VRPM (Transverse-Flux) Machines. In Proceedings of the 8th International Conference on Electrical

Machines and Drives, Cambridge, UK, 1–3 September 1997; pp. 386–390.

9 Bang, D.-J.; Polinder, H.; Shrestha, G.; Ferreira, J.A. Comparative Design of Radial and Transverse Flux PM Generators for Direct-Drive Wind Turbines. In Proceedings of the 18th International Conference on Electrical Machines, Vilamoura, Portugal, 6–9 September 2008; pp. 1–6.

10 Siatkowski, M.; Orlik, B. Flux Linkage in Transverse Flux Machines with Flux Concentration. In Proceedings of the International Conference on Optimization of Electrical and Electronic Equipment (OPTIM 2008), Brasov, Romania, 22–24 May 2008; pp. 21–26.

11 Qin, D.; Qu, R.; Lipo, T.A. A Novel Electric Machine Employing Torque Magnification and Flux Concentration Effects. In Proceedings of the 1999 IEEE Thirty-Fourth IAS Annual Meeting, Industry Applications Conference, Phoenix, AZ, USA, 3–7 October 1999; Volume 1, pp. 132–139.

12 Harris, M.R.; Pajooman, G.H.; Sharkh, S.M.A. Comparison of Alternative Topologies for VRPM (Transverse-Flux) Electrical Machines. In Proceedings of the IEE Colloquium on New Topologies for Permanent Magnet Machines, London, UK, 18 June 1997; pp. 2/1–2/7.

13 Siatkowski, M.; Orlik, B. Influence of Saturation Effects in a Transverse Flux Machine. In Proceedings of the 13th Power Electronics and Motion Control Conference (EPE-PEMC 2008), Poznan, Poland, 1–3 September 2008; pp. 830–836.

14 Zhao, Y.; Chai, J. Power Factor Analysis of Transverse Flux Permanent Machines. In Proceedings of the 8th International Conference on Electrical Machines and Systems (ICEMS 2005),

Nanjing, China, 27–29 September 2005; Volume 1, pp. 450–453.

15 Guo, Y.G.; Zhu, J.G.; Watterson, P.A.; Wu, W. Comparative study of 3-D flux electrical machines with soft magnetic circuit. *IEEE Trans. Ind. Appl.* **2003**, *39*, 1696–1703.

16 Svechkarenko, D.; Soulard, J.; Sadarangani, C. Parametric Study of a Transverse Flux Wind Generator at No-Load Using Three-Dimensional Finite Element Analysis. In Proceedings of the International Conference on Electrical Machines and Systems (ICEMS 2009), Tokyo, Japan, 15–18 November 2009; pp. 1–6.

17 Masmoudi, A.; Elantably, A. A Simple Assessment of the Cogging Torque in a Transverse Flux Permanent Magnet Machine. In Proceedings of the IEEE International Electric Machines and Drives Conference (IEMDC 2001), Cambridge, USA, 17–20 June 2001; pp. 754–759.

CHAPTER 5

Research on the Torque and Back EMF Performance of a High Speed PMSM Used for Flywheel Energy Storage

Jing Zhao, Zhongxin Gu *, Bin Li, Xiangdong Liu, Xiaobei Li and Zhen Chen

School of Automation, Beijing Institute of Technology, Beijing 100081, China

ABSTRACT

Due to advantages such as high energy density, high power density, rapid charge and discharge, high cyclic-life, and environmentally friendly, flywheel energy storage systems (FESs) are widely used in various fields. However, the performance of FES systems depends on the performance of a high speed machine, therefore, the design and optimization of a high efficiency and high power density machine are very crucial to improve the performance of the whole FES system. In this paper, a high speed permanent-magnet synchronous machine (PMSM) is researched. Considering the requirement of low torque ripple in low speed and loss caused by back electromotive force (EMF) harmonics, the electromagnetic performance is improved from points of view of slot/pole matching, magnetic-pole embrace with the finite element method (FEM). Furthermore, the magnetic-pole eccentricity, the slot opening, the thickness of PM and air-gap length are also optimized with Taguchi method. The electromagnetic performance, such as torque ripple, cogging torque, average torque and back EMF wave are much improved after optimization. Finally, experiments are carried out to verify the calculated results.

KEYWORDS

flywheel energy storage (FES); high speed; permanent-magnet synchronous machine (PMSM); torque ripple; Taguchi method; back electromotive force (EMF); cogging torque

1. INTRODUCTION

With the rapid development of society and the economy, higher requirements are put forward for energy use, which include more reliability, more flexibility, more efficiency, more adaptability to various actual application situations, *etc.*Energy storage is a preferable way to solve the energy crisis problem and enable a sustainable energy future [1,2]. There are five main energy storage technologies, including storage batteries, super-capacitors, pumped storage, compressed air energy storage (CAES) and flywheel energy storage (FES) [3,4]. Table 1 shows a comparison between the five storage technologies [5,6,7,8]. As is seen from the Table, it is obvious that FES has the advantages of high power density, high storage density, rapid charge and discharge, easy installation and maintenance, high cyclic-life, and being environmentally friendly. Therefore, the FES has broad application prospects.

Table 1. Comparison of five storage technologies.

Index	Super-Capacitor	CAES	Pumped Storage	Storage Battery	FES
Efficiency (%)	~90	<50	~60	~70	~90
Storage density	High	High	-	Middle	Highest
Power density	High	Middle	-	Middle	Highest
Modularity	No	No	No	Yes	Yes
Service life(years)	10	5	10	5	20
Charge time unit	Hour	Hour	Hour	Hour	Minute
Pollution	No	Yes	Yes	Yes	No
Build cycle unit	Year	Year	Year	Month	Week

In recent years, the process of application of FES has been accelerated by the development of power electronic technology, motor/generator integrated technology, control technology, bearing technology and materials technology [9,10,11]. FESs are widely applied in various fields, such as power system applications, transportation applications, pulsed power applications, uninterrupted power supply (UPS) applications, space applications and military applications, *etc.* As for power system applications, the FES can be used for power regulation between utilities in the power grid [12,13,14,15,16]. For space applications, the FES can supply energy storage with the smallest mass and also can be used for attitude control for guidance of the craft [17,18,19]. For UPS applications, the FES system can serve as an UPS to supply energy for vital loads, such as hospitals, central computer rooms, government agencies, *etc.* [20,21,22]. As for military applications, the FES system can supply short high power bursts for electromagnetic cannon [23]. The aforementioned applications of FES show that FES plays an important role in modern industry and economy.

As shown in Figure 1a, a FES system mainly consists of four parts: flywheel, motor/generator, power electronic converter and controller [15,17]. The flywheel is coaxially connected with the shaft of the motor/generator. When the motor/generator operates as a motor, the system is charged and electric energy is transformed into mechanical energy by the flywheel. When electric energy is needed, the motor/generator will operate as a generator and the mechanical energy stored in the flywheel will be transformed into electric energy through power electronic converter. The drive circuit is shown in Figure 1b. As shown in this figure, the left and right rectifier/inverters that are controlled by the drive circuit controller work as rectifier and inverter, respectively, during the charging stage, and the power flows to the load and flywheel. During the discharging stage, the right rectifier/inverter works as a rectifier, while the left rectifier/inverter can work as a rectifier or an inverter depending on the load requirements. As is known, a multilevel converter can reduce the harmonics in the output voltage. Besides the converter topology shown in Figure 1b, the circuit topologies of diode-clamped multilevel inverter, capacitor-clamped multilevel inverter and cascaded inverter can be used in a FES [24]. Furthermore, modulation strategies, such as sinusoidal pulse-width modulation (SPWM) and space-vector modulation (SVM) can be used to obtain more sinusoidal output voltage.

The FES system stores kinetic energy when it works. The amount of stored energy E_k is defined by the following mechanical relationship:

$$E_k = \frac{1}{2} J\omega^2 \tag{1}$$

where J is the inertia of the flywheel and rotor, ω is the rotational speed of flywheel or rotor. It is obvious that high rotational speed of the rotor can increase the energy density. The performance of the high speed machine has much influence on the FES system. So this paper is written from aspect of machine design.

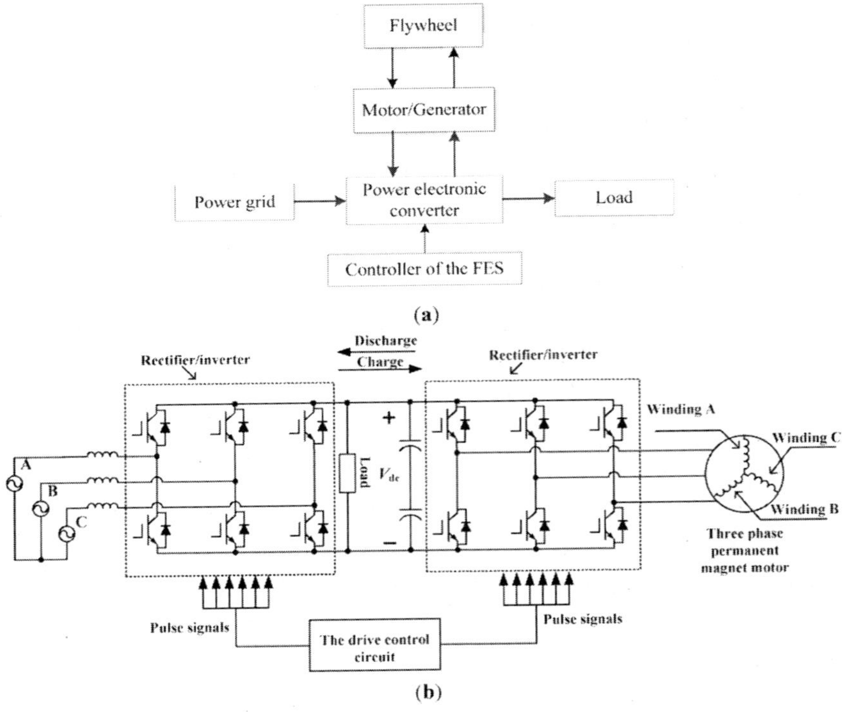

Figure 1. (**a**) System block diagram of the FES; (**b**) drive circuit.

The permanent-magnet synchronous machine (PMSM) is favorable for use as a motor/generator in FES, because the PMSM has the advantages of high efficiency, high power density and high power factor, and so on. However, a lot of attention has been focused on the loss calculation and thermal analysis of the high speed machine [25,26,27,28,29,30,31,32,33]; in [25,26,27], a calculation method for the harmonic iron loss based on the combination of 2D and 3D FEMs is proposed; the iron loss and copper loss of a high speed PMSM that can operate at a speed of 60,000 rpm are analyzed and calculated in [28]; in [29], a fast hybrid method considering the factors of end effect and pulse-width modulation harmonics is proposed to calculate the iron loss; loss and thermal characteristics of a high speed PMSM based on the FEM are analyzed in [30,31,32]; in [33], a combined network and computational fluid dynamic (CFD) method is used for the thermal analysis. However, considering that the high speed machine used for FES system should start easily and run stably, the machine should have good torque performance. In addition, considering that FES system should have low iron loss at no load, air-gap magnetic field harmonics, which can be reflected by the back electromotive force (EMF) harmonics, will cause high losses in the stator core, rotor core and permanent magnets at high speed [34,35,36,37]. Furthermore, when the FES works under discharging or charging conditions, the back EMF waveform must also be closer to a sine wave and have as few harmonics as possible. Thus the torque performance and the back EMF have received much attention in this paper. The rotor structure can influence the back EMF and torque performance of a high speed machine. Ref. [28] adopted surface PMs while ref. [27] adopted interior PMs. This paper will compare the surface PMs with interior PMs from aspects of torque performance and back EMF harmonics. Numerous methods have been researched for reducing cogging torque, torque ripple and back EMF harmonics to optimize the design of PMSMs [38,39,40,41,42,43,44,45,46,47,48,49,50,51].

A fractional number of slots per pole is an effective method. The larger the least common multiple of slots and poles is, the smaller the cogging torque is [38,39]. However, these works don't consider the influence on the back EMF when the pole number is eight or above. In a high speed machine, a large pole number will cause high iron losses, so this paper focuses on the selection of slot number considering both torque performance and back EMF when the pole number has a constant value of four. The skewing can also effectively reduce both cogging torque and back EMF harmonics. The skewing can be designed at the stator, rotor or magnet [40,41]. However, compared with other methods, skewing has high manufacturing technology requirements, which will consequently increase the manufacturing cost. In addition, magnet shape has great influence on the torque performance and back EMF. Adopting a proper magnetic-pole embrace, cogging torque can reach zero in theory for surface machines and the back EMF harmonics can also be decreased. Bread-shape and stacked PMs are adopted to decrease the torque ripple in [42], but the torque ripple has been reduced at the expense of high manufacturing cost and the stacked PMs have the risk of falling off when the machine runs at high speed. Ref. [43] and ref. [44] only considered the back EFM wave, while ref. [45] only considered the cogging torque reduction. Ref. [46] optimized the torque performance considering both magnetic-pole embrace and magnetic bridge width, and the magnetic-pole embrace of all poles is the same. This paper considers both torque performance and the back EMF during optimization. The N poles and S poles of the high speed machine that are set with different magnetic-pole embrace values will be researched. Sometimes the stator structure, such as teeth shape, is also optimized to minimize the cogging torque [47,48]. Besides, calculated optimization methods have attracted much attention in recent years. In [49], a genetic algorithm is used to optimize the discrete skew angle of PMs to reduce cogging torque based on 3D FEM. However, this work just optimized one parameter that has an influence on the torque performance and there are no experiments to verify the optimization result. In [50], three motor parameters are optimized at the same time to reduce the cogging torque by the non-dominated sorting genetic algorithm, but the optimization step length of each parameter is very small, which requires a large calculation time. The most important point is that it is impossible for a motor designer to use a complicated and time-consuming algorithm during the practical motor design. The Taguchi method [51,52] is a less time-consuming and multi-parameter method. Ref. [52] performed a robust design of EMF to reduce the effect caused by the manufacturing tolerances, and width, thickness and residual of PM are selected as the noise variables. In this paper, the magnetic-pole eccentricity, the slot opening, the thickness of PM and the length of air gap are optimized by using the Taguchi method for machine design. The tradeoff between back EMF and torque performance is considered. Experiments are conducted to verify the calculated results. This paper can provide useful references for a high speed machine designer.

This paper almost considers most of the factors that have an influence on the torque and back EMF performance. Meanwhile, a simple and effective optimization method, such as magnetic-pole embrace, is used for preliminary optimization, and a less time-consuming and multi-parameter method is adopted for further optimization. Sine-wave current is adopted to decrease the torque ripple and the harmonics of the back EMF wave. This paper is organized as follows: a rotor structure for the high speed machine is selected and the initial machine parameters are obtained in Section 2; based on the initial scheme, comprehensive analysis and optimizations are carried out by selecting an optimal slot/pole combination and magnetic-pole embrace with the finite element method (FEM) in Section 3; partial parameters, such as the magnetic-pole eccentricity, the slot opening, the thickness of PM and the length of air gap, are also optimized by using the Taguchi method in Section 4; a prototype high speed PMSM was manufactured and tested to validate the simulation results in Section 5.

2. SELECTION OF ROTOR STRUCTURE

Both of the rotors with surface PMs (SPMs) and interior PMs (IPMs) can be used in high speed PMSMs. Both structures both adopt 12 slots-4 poles and the same stator parameters. Their rotor structures are different, but with the same volume of magnet materials, as shown in Figure 2.

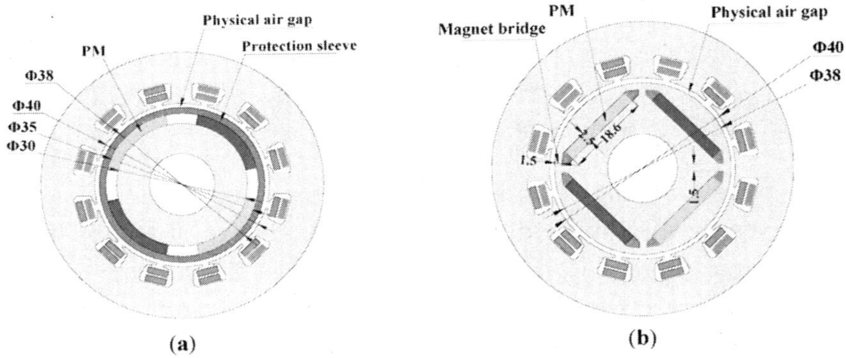

Figure 2. Dimensions of the two rotors: (**a**) SPM and (**b**) IPM.

As for surface PMSM, a non-magnetic protection sleeve is usually adopted to protect the PMs from the high centrifugal force. In this paper, the total equivalent air gap length of the surface PMSM is 2.5 mm which includes the thickness of the protection sleeve (1.5 mm) and the physical air gap length (1 mm). Compared with the surface PMSM, the equivalent air gap length of an interior PMSM can be decreased to 1 mm without the protection sleeve. Besides, the magnetic bridge thickness of the interior PMSM is set to 1.5 mm to ensure the same mechanical strength as the surface PMSM. The PM width and PM thickness of the interior PMSM are 18.6 and 2.5 mm, respectively.

The surface PMSM and interior PMSM are compared by the finite element method (FEM), and the results are summarized in Table 2. It can be seen that although the interior PMSM has a smaller equivalent air gap length and the contribution of reluctance torque, it has lower average torque (T_{av}). In addition, the interior PMSM has higher cogging torque (T_{cog}), higher torque ripple (T_{rip}) and higher total harmonics distortions (THD) than the surface PMSM. This is because the magnetic flux leakage (leakage coefficient of 1.6) of the interior PMSM is much higher than that of surface PMSM (leakage coefficient of 1.13) due to the large magnetic bridge thickness, as shown in Figure 3. Thus, the rotor structure with surface PMs is chosen for further research. The initial parameters of the PMSM with SPM rotor are listed in Table 3.

Table 2. Performances Comparison of IPM and SPM machines.

Structures	T_{av}	T_{rip}	T_{cog}	THD
IPM	385 mN·m	6.60%	27.6 mN·m	11.50%
SPM	387 mN·m	3.20%	21.4 mN·m	7.30%

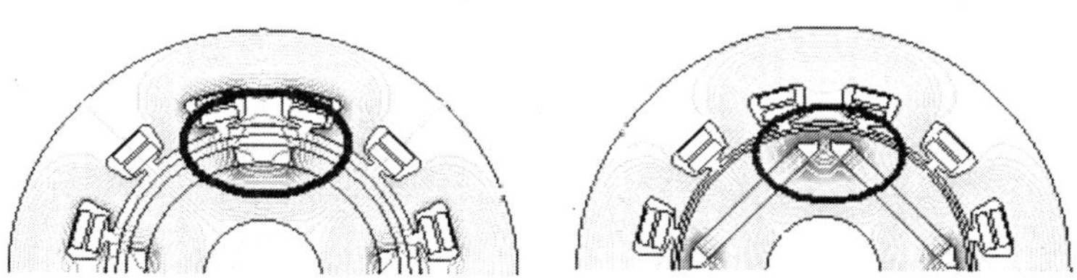

(a) (b)

Figure 3. Flux line distribution of the two structures: (**a**) SPM and (**b**) IPM.

Table 3. Initial parameters of the machine.

Parameters	Value	Parameters	Value
Magnet material	NTP42UH	Magnet remanence (20 °C) (T)	1.299
Pole number	4	Rated speed (rpm)	30000
Rated power (kW)	1	Conductor per slot	24
Length of core (mm)	80	Outer diameter of rotor (mm)	35
Outer diameter of stator (mm)	65	Inner diameter of rotor (mm)	15
Inner diameter of stator (mm)	40	Magnetic-pole embrace	0.73
Thickness of magnet (mm)	2.5	Magnetic-pole eccentricity (mm)	0
Thickness of protection sleeve (mm)	1.5	Slot opening (mm)	1.5
Current density (A/mm^2)	6.4	Air gap length (mm)	1
Linear current density (A/cm)	50	Average air-gap flux density (T)	0.636

3. OPTIMIZATION OF POLE/SLOT MATCHING AND MAGNETIC POLE EMBRACE

3.1. Pole/Slot Combination

Cogging torque is an important factor to torque characteristics. The cogging torque can be calculated as [53]:

$$T_{cog}(\alpha)=-\frac{L_{ef}B_{\sigma}^{2}C_{T}}{\mu_0\pi}\left(R_2^2-R_1^2\right)\times\sum_{n=1}^{\infty}\frac{K_{skn}}{n}\sin(nN_L\frac{b_0}{2})\sin(nN_L\frac{\alpha_p\pi}{N_p})\sin(nN_L\alpha-\frac{1}{2}nN_L\alpha_s) \quad (2)$$

where μ_0 is the permeability of air, L_{ef} represents the effective axial length, B_δ represents magnetic flux density of air gap, R_1 and R_2 represent outer and inner diameter of air gap, N_p represents pole number, N_L represents the least common multiple (LCM) between N_p and N_s, b_0 represents the angle of slot opening, α_p represents the magnetic-pole embrace, $\alpha_s$represents the angle of stator skew, α represents the rotor position, K_{skn} is the skewing factor, and C_T is given by:

$$C_T = N_s N_p / N_L \qquad (3)$$

where N_s represents slot number.

It can be seen from Equations (2) and (3) that the slot/pole combination can affect the coefficient C_T which will further affect the cogging torque. Therefore, it is necessary to analyze the slot/pole combination to obtain low torque ripple. In this paper, three high speed PMSMs with different slot/pole combinations of 4 poles-12 slots (4p-12s), 4 poles-18 slots (4p-18s) and 4 poles-24 slots (4p-24s) are researched and compared. The three motors differ in the number of slots and conductors per slot, but they have the same number of series turns per phase, outer diameter of stator and other parameters as listed inTable 3. The torque characteristics results are shown in Table 4.

Table 4. Performances of machine with different slot-pole combination.

Slot/Pole Combination	4p-12s	4p-18s	4p-24s
T_{cog} (mN·m)	21.3	4.1	7.0
T_{av} (mN·m)	387.9	421.4	413.9
T_{rip}	3.2%	1.6%	1%
THD	7.3%	4.1%	3.6%

Based on the data, the 4p-12s type is abandoned because of its lowest torque (387.9 mN·m) and highest torque ripple (3.2%). The 4p-18s has a higher average torque and higher torque ripple than the 4p-24s version, while the cogging torque is smaller than that of the 4p-24s type. The torque ripple contains cogging torque and electromagnetic pulsation, which has relationships with the THD of EMF. The phenomenon that the 4p-18s has a smaller cogging ripple but a higher torque ripple can be explained by the higher THD of the 4p-18s (4.1%) than that of the 4p-24s (3.6%). Considering all these factors, it is difficult to select a better structure between 4p-18s and 4p-24s, therefore, further analysis is needed.

3.2. Optimization of Traditional Magnetic-Pole Embrace

As to PMSM, both the amplitude and distribution of the air-gap magnetic field will be affected by changing of the magnetic-pole embrace, which will further influence the torque characteristics. It can be seen from Equation (2) that the magnetic-pole embrace also has an influence on cogging torque. It can be analyzed that the cogging torque will be zero in theory if the magnetic-pole embrace satisfies the flowing equation:

$$\sin(nN_{\mathrm{L}}\frac{\alpha_{\mathrm{p}}\pi}{N_{\mathrm{P}}})=0 \tag{4}$$

The optimal theoretical magnetic-pole embrace will be calculated and shown as:

$$\alpha_{\mathrm{p}}=\frac{N_{\mathrm{L}}/N_{\mathrm{P}}-k}{N_{\mathrm{L}}/N_{\mathrm{P}}} \tag{5}$$

where k is positive integer below the value of N_L/N_p.

Therefore there should be some magnetic-pole embraces making the cogging torque and torque ripple the smallest. In order to ensure the magnetic flux per pole and the average torque, the magnetic-pole embrace is varied from 0.6 to 1.0. The trends of average torque and cogging torque of both 4p-18s PMSM and 4p-24s PMSM are calculated by FEM, as shown inFigure 4 and Figure 5. It can be seen that with the increase of magnetic-pole embrace, average torque continues increasing. In addition, the 4p-18s has a higher average torque for the same magnetic-pole embrace. However, the cogging torque fluctuates with the increasing magnetic-pole embrace. The theoretical and simulated optimal values of α_p are calculated and summarized in Table 5. As is seen from Figure 5 and Table 5, the 4p-18s has three optimal points (0.71 0.81 and 0.92) and 4p-24s has two optimal points (0.72 and 0.88) to minimize the cogging torque, which is consistent with the theoretical analysis. It also can be seen that the simulated optimal magnetic-pole embrace values are a little higher than the theoretical values, which is due of the neglect of magnetic flux leakage in Equation (5).

The torque characteristics at the optimal points are summarized in Table 6. In order to obtain a better characteristic, two optimal points of 0.81 and 0.92 for 4p-18s PMSM and two optimal points of 0.72 and 0.88 for 4p-24s PMSM are selected for further research in the following section.

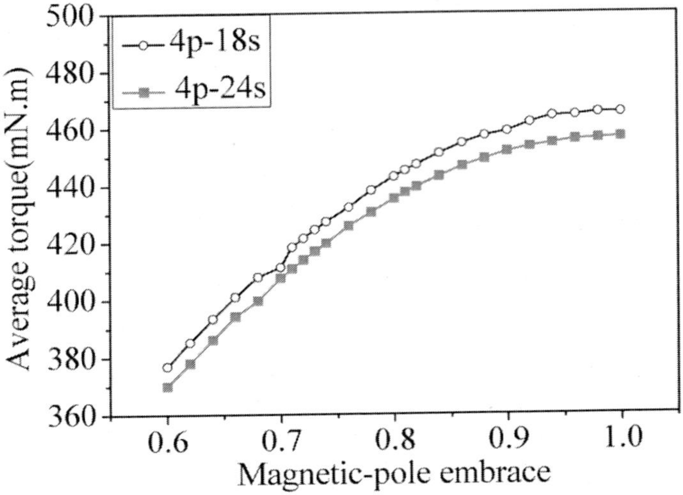

Figure 4. The trend of average torque with the magnetic-pole embrace increasing.

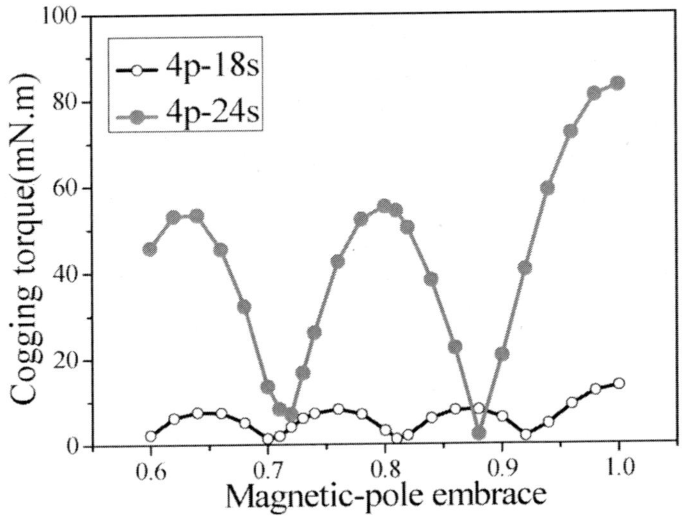

Figure 5. Influence of the magnetic-pole embrace on the cogging torque.

Table 5. The Optimal Magnetic-Pole Embrace.

Slot/Pole Combination	4p-18s	4p-24s
N_L/N_p	9	6
Theoretical Value with fraction form	8/9,7/9,2/3 (0.89,0.78,0.67)	5/6,2/3 (0.83,0.67)
Simulated Optimal Value	0.92,0.81, 0.71	0.88,0.72
Quantitative Difference	0.03,0.03,0.04	0.05,0.03

Table 6. The torque characteristics at simulated optimal magnetic-pole embrace.

Slot/Pole Combination	4p-18s			4p-24s	
Optimal Magnetic Pole Embrace	0.71	0.81	0.92	0.72	0.88
Average Torque (m·Nm)	418.3	445.0	461.8	413.9	449.0
Torque Ripple (%)	1.4	0.2	1.5	1.0	0.6
Cogging Torque (m·Nm)	2.0	1.2	1.8	7.0	2.3

3.3. Optimization of the Torque Characteristic with a Magnetic-Pole Embrace Combination

The machine may have better torque characteristics when the N poles and S poles of the machine are set with different magnetic-pole embrace values, which is shown in Figure 6. In this part, the combination of the N-pole and S-pole magnetic-pole embraces is optimized around the optimal magnetic-pole embraces obtained. In this process, all the N poles have the same magnetic-pole embrace and all the S poles have another magnetic-pole embrace. As for 4p-18s, the magnetic-pole embrace of the S poles varies from 0.82 to 0.96, while that of the N poles varies from 0.77 to 0.85; the corresponding cogging torques are shown in Figure 7a. As for 4p-24s, similarly, the magnetic-pole embrace of the S poles varies from 0.84 to 0.96 and that of the N poles from 0.68 to 0.76, and the corresponding cogging torques are shown in Figure 7b.

From Figure 7, the points which have the smallest cogging torque are selected and summarized in Table 7. In addition, the torque characteristics at these points are calculated and listed in Table 7 and the torque characteristics at simulated optimal magnetic-pole embrace from Table 6 are also listed in the Table 7.

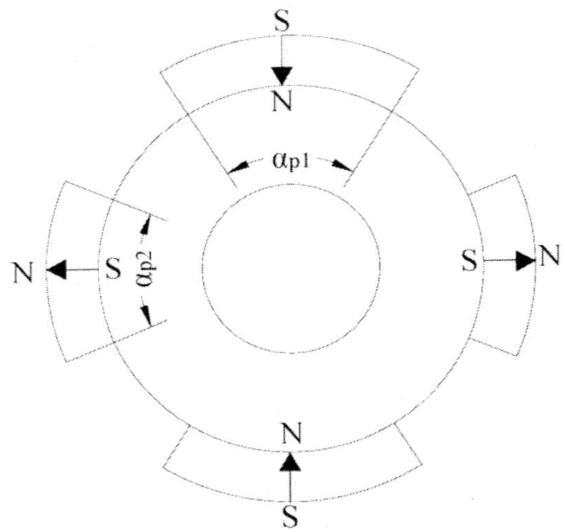

Figure 6. N poles and S poles of the machine with different magnetic-pole embrace values.

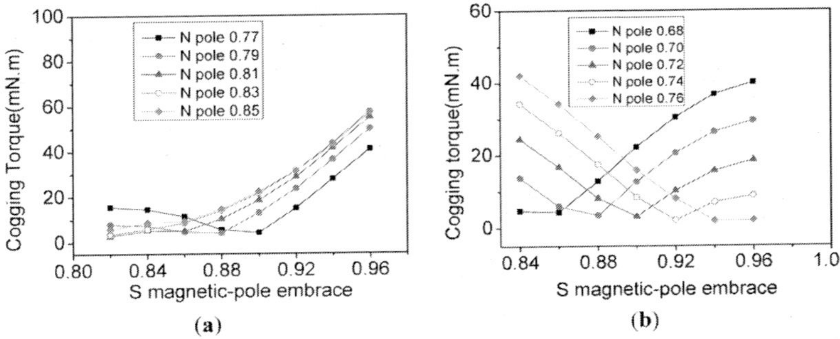

Figure 7. The trend of cogging torque with different magnetic-pole embrace: (**a**) 4p-18s PMSM; (**b**) 4p-24s PMSM.

Table 7. Torque characteristics with combinations of embrace.

Slot/Pole Combination	No.	N Poles	S Poles	T_{cog} (mN·m)	T_{av} (mN·m)	D_{rav} (%)	T_{rip} (%)	D_{rrip} (%)
4p-18s	1	0.77	0.9	4.3	447.3	3.1	1.49	0
	2	0.79	0.88	4.2	448.2	2.9	1.10	26.2
	3	0.81	0.82	2.9	446.1	3.4	0.37	75.2
	4	0.83	0.82	3.6	448.1	3.0	0.59	60.4
	5	0.85	0.82	5.9	450.1	2.5	0.95	36.2
	6	0.81	0.81	1.2	445.0	3.6	0.19	87.3
	7 (as reference)	0.92	0.92	1.8	461.8	0.0	1.49	0.0
4p-24s	1	0.68	0.86	4.3	422.7	8.5	0.53	64.4
	2	0.7	0.88	3.4	427.8	7.4	0.42	71.8
	3	0.72	0.9	3.1	431.6	6.5	0.41	72.5
	4	0.74	0.92	2.0	439.3	4.9	0.38	74.5
	5	0.76	0.94	1.9	439.6	4.8	0.41	72.5
	6	0.72	0.72	7.0	413.9	10.44	0.97	34.9
	7	0.88	0.88	2.3	449.0	2.8	0.58	61.1

After the combinations, the 4p-24s PMSM with different magnetic-pole embrace gives a smaller cogging torque. For example, when the N and S poles have a magnetic-pole embrace of 0.76 and 0.94, respectively, the cogging torque is 1.9 mN·m and torque ripple is smaller (0.41%), but it is difficult to select an optimal scheme among these schemes directly. In order to evaluate torque characteristic more objectively, the decreasing rates of average torque (D_{rav}) and torque ripple (D_{rrip}) are calculated by taking the values in the No. 7 combination of the 4p-18s PMSM as the reference. A PMSM with the same magnetic-pole embrace 0.81 (No. 6 combination) is selected as the best scheme based on the compromise principle: the torque ripple and cogging torque are effectively reduced, but with the average torque loss being as little as possible.

4. PERFORMANCE OPTIMIZATION WITH THE TAGUCHI METHOD

As is known, a high frequency of the fundamental and especially the harmonic magnetic field in a high speed machine will cause high losses in stator core, rotor core and permanent magnets. Considering that the back EMF waveform is the reflection of the magnetic field distribution, the THD of back EMF is studied. After the optimization in Section 3, the machine has good torque characteristics with an average torque of 445 mN·m and a torque ripple of 0.19%. However, the THD of back EMF is 9.72%, which is still high and will cause high losses. Thus a further THD optimization is needed based on the scheme obtained in Section 3.

In traditional methods, if there are N variables influencing EMF harmonics, average torque and the cogging torque, it is hard to take them into consideration together because of the complicated combinations. Here we give an example: if each variable has three levels of setting, there may be 3^N design variations (N is the number of the variable) in the traditional method. The Taguchi method is a design optimization method. It can simplify the design process, reduce the calculation times, and provide a systematic and efficient approach to determine the optimum settings of multiple design parameters.

As shown in Figure 8, there are four main factors influencing the back EMF and torque characteristic: the magnetic-pole eccentricity (D_{off}), the slot opening (b_{s0}), the thickness of PM (T_{PM}) and the equivalent length of air gap (δ). The D_{off} is obtained by fixing O_1 and moving O_2. The minimum air gap is unchanged as D_{off} varying. The base value of the four factors are $D_{\mathrm{off}} = 0$, $b_{s0} = 1.5$ mm, $T_{\mathrm{PM}} = 2.5$ mm, $\delta = 2.5$ mm, respectively. They will be optimized with the Taguchi method.

4.1. Optimization with Taguchi Method

There are three stages to perform the Taguchi method: (1) First of all, four factors A, B, C and D corresponding to D_{off}, b_{s0}, T_{PM} and δ

are selected, the different levels of all factors are shown in Table 8; (2) Secondly, as shown in Table 9, the four factors at three levels are combined based on the orthogonal array (a standard OA L_9), which is the most efficient orthogonal design.

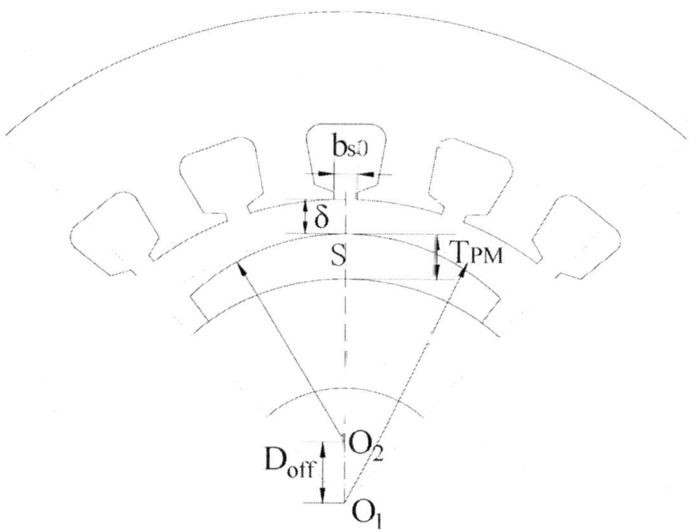

Figure 8. Four parameters to be optimized: D_{off}, b_{s0}, T_{PM}, δ.

Table 8. Levels of design variables.

Design Variable (mm)	Level 1	Level 2	Level 3
A: D_{off}	0	2	4
B: b_{s0}	1.0	1.5	2
C: T_{PM}	2	2.5	3
D: δ	2.3	2.5	2.7

Table 9. Results of nine models.

No.	A	B	C	D	T_{cog} (mN·m)	T_{rip}	T_{av} (mN·m)	THD
1	0	1.0	2	2.3	1.4	0.28%	429.0	10.4%
2	0	1.5	2.5	2.5	1.1	0.19%	445.0	9.72%
3	0	2.0	3	2.7	2.1	0.25%	450.7	9.01%
4	2	1.0	2.5	2.7	0.6	0.18%	406.2	6.52%
5	2	1.5	3	2.3	2	0.28%	481.0	7.05%
6	2	2.0	2	2.5	1.2	0.29%	375.5	6.41%
7	4	1.0	3	2.5	0.6	0.25%	437.1	3.47%
8	4	1.5	2	2.7	0.1	0.28%	331.3	1.48%
9	4	2.0	2.5	2.3	0.6	0.31%	415.4	2.94%

The corresponding results of torque characteristics and the THD of back EMF for each combination scheme are calculated with FEM and also listed in Table 9; (3) Finally, the calculated results in Table 9 are analyzed through analysis of means (ANOM) and analysis of variance (ANOVA) to obtain optimal setting of parameters [54], which will be carried out in the following part.

4.2. Analysis of Results

4.2.1. Analysis of Means (ANOM)

ANOM is carried out to determine the best level combination to achieve optimal performance. From Table 9, it can be seen that all the cogging torque and torque ripple are very small, which can meet the requirements. Thus, the value of THD of back EMF and average torque will be emphatically analyzed in this part. The means of all results can be calculated as:

$$m=\frac{1}{9}\sum_{i=1}^{9} T(i) \tag{6}$$

When the average effect of a design variable at one setting is considered, the average value of the design variable for one level of the setting factor can be calculated. For example, the average THD of setting factor B at level 2 (b_{s0} = 1.5 mm), which is only at No. 2, 5 and 8 in Table 9, is calculated as:

$$m_{B_2} = \frac{1}{3}\sum_{i=1}^{3} T_{THD}(2) + T_{THD}(5) + T_{THD}(8) \tag{7}$$

Average THD and average torque for all levels of all factors can be obtained in a similar way. The THD and average torque result data, which is called the main factor effects, are listed in Table 10 and Table 11, respectively. As shown inTable 10 and Table 11, it is clear that the factor-level combination of A3-B2-C1-D3 (D_{off} = 4 mm, b_{s0} = 1.5 mm, T_{PM} = 2 mm, δ = 2.7 mm) contributes to the minimization of THD, and the factor-level combination of A1-B1-C3-D1 (D_{off} = 0 mm, b_{s0} = 1.0 mm, T_{PM} = 3 mm, δ = 2.3 mm) contributes to the maximization of average torque. In order to determine the factors that have significant effects on the machine performance, a mathematical method called ANOVA is used in the following section to obtain the ultimate optimal scheme for the high speed PMSM.

Table 10. THD for all levels of all factors.

Setting Factors	A_i	B_i	C_i	D_i
i = 1	9.70%	6.80%	6.09%	6.79%
i = 2	6.70%	6.08%	6.39%	6.53%
i = 3	2.60%	6.12%	6.51%	5.67%

Table 11. Average torque for all levels of all factors.

Setting Factors	A_i	B_i	C_i	D_i
$i = 1$	441.6	424.1	378.6	441.8
$i = 2$	421	419.1	422.2	419.2
$i = 3$	394.6	413.9	456.3	396.1

4.2.2. Analysis of Variance

ANOVA does not analyze the results directly, but rather determines the relative importance of various design variables. In order to perform ANOVA, the sum of squares (SS), which is calculated based on the data of ANOM, should be obtained firstly. The SS for factor B can be calculated as:

$$SS_{FB} = 3\sum_{i=1}^{3}\left(m_{B_i} - m\right)^2 \tag{8}$$

SS_F for A, C and D are obtained in the similar way. All the SS data for THD and average torque are summarized in Table 12. The effect ratio for each variable is calculated and also summarized in Table 12.

As shown in Table 12, the factors A (D_{off}) and B (b_{s0}) have larger effects on THD than average torque, while the factor C (T_{PM}) and the factor D (δ) have larger effects on average torque than THD. Thus, based on the selected combination A3-B2-C1-D3 for THD and combination A1-B1-C3-D1 for average torque, the best combination of A3-B2-C3-D1 (D_{off} = 4 mm, b_{s0} = 1.5 mm, T_{PM} = 3 mm, δ = 2.3 mm) is obtained to reduce the THD and increase the average torque as much as possible.

Table 12. Effects of various factors on machine performance indexes.

Factors	THD		Average Torque	
	SS_F	Factor Effect Ratio	SS_F	Factor Effect Ratio
A	76.2	95.8%	3330.4	21.2%
B	1	1.2%	156.1	1%
C	0.3	0.4%	9101.1	57.9%
D	2.1	2.6%	3132.9	19.9%
Total	79.6	100%	15720.5	100%

The performances of the initial and the final schemes are compared in Figure 9 and Table 13. After the final improvement, THD of the back EMF waveform decreases from 7.30% to 3.20% (by 50.3%), while cogging torque decreases from 21.36 to 1.04 mN·m (by 95.1%) and torque ripple decreases from 3.63% to 0.29% (by 90.9%). Meanwhile, the average torque increases from 387 to 459.2 mN·m (by 17.9%). It is validated that the electromagnetic performances of the machine are effectively improved. Furthermore, the rated *q*-axis reactance (X_q = 3.3938 Ω) and the rated *d*-axis reactance (X_d = 3.4263 Ω) are obtained using methods of back EMF by the FEM [55].

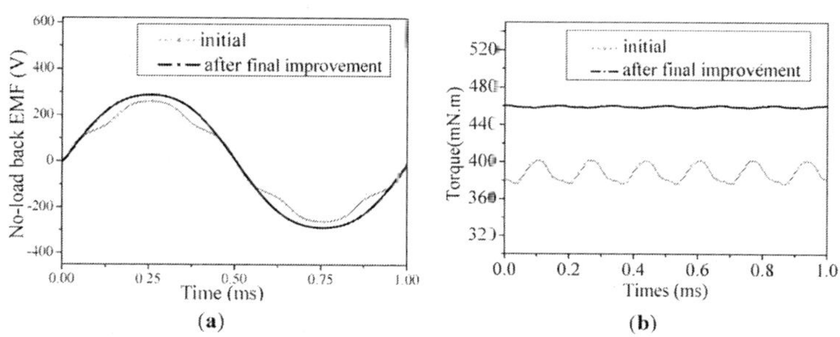

Figure 9. Comparison of electromagnetic performance of before and after optimization: **(a)** no-load back EMF waveform; **(b)** torque waveform.

Table 13. Comparison results.

Performances	Initial	Final	Increasing Rate
T_{av} (mN·m)	387	456.3	17.9%
THD of back EMF	7.30%	3.63%	−50.3%
T_{cog} (mN·m)	21.36	1.04	−95.1%
T_{rip}	3.20%	0.29%	−90.9%

5. PROTOTYPE MACHINE AND EXPERIMENTS

Based on the optimized scheme of the high speed PMSM, a prototype machine was manufactured, as shown in Figure 10. In order to evaluate the performance of the prototype machine, the no-load back EMF and torque characteristic are tested.

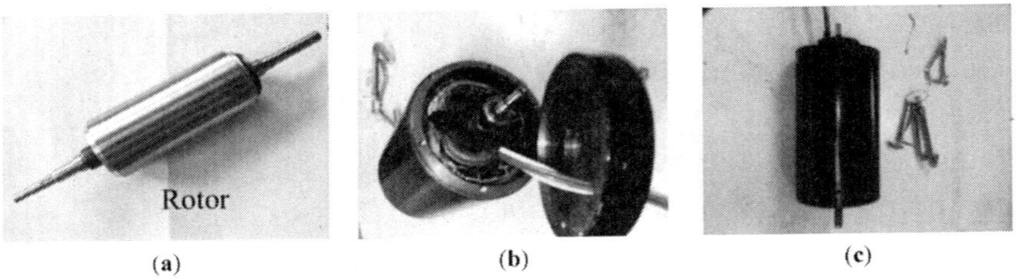

Figure 10. High speed PMSM prototype: (**a**) Rotor; (**b**) Stator; (**c**) Prototype.

5.1. No-Load Back EMF Test

Because of the lack of another high speed machine as the driving machine, the no-load back EMF test was limited to less than 5000 rpm. The measured and calculated no-load back EMFs at different speeds are shown in Figure 11. As can be seen from this figure, the calculated and measured back EMFs are in good agreement, which validates the results simulated by FEM. It can be seen from Figure 11b that the peak-peak value of the no-load back EMFs increases linearly with the speed increasing. Figure 11a shows the calculated and measured no-load back EMFs at 4860 rpm, and they are approximately sinusoidal. Especially, the THD of the measured back EMF is 2.59%, which is even smaller than the designed results. These validate the good design of the magnetic structure.

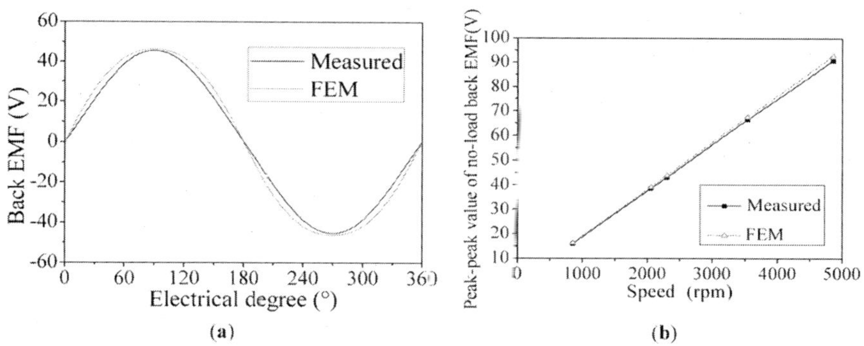

Figure 11. No-load Back EMF (**a**) at 4860 rpm and (**b**) *versus* different speed.

5.2. Torque Test

The torque characteristic at load is tested by a high- precision speed/torque sensor at very low speed, as shown in Figure 12 and Figure 13.

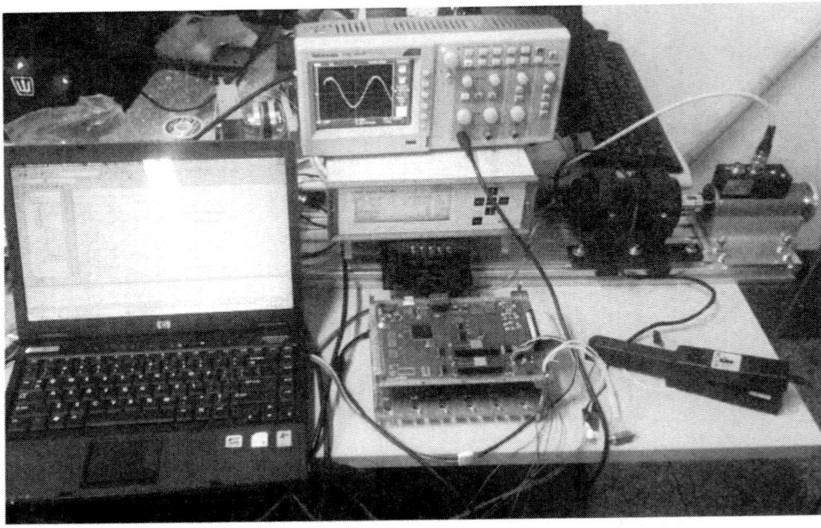

Figure 12. Test bench for measuring torque at load.

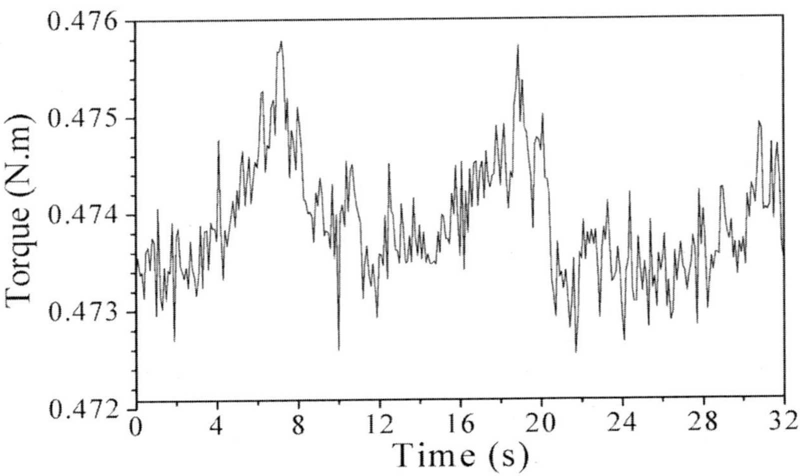

Figure 13. Measured torque at load.

The measured average torque and torque ripple at load are 473.9 mN·m and 0.34%, respectively. Because the load is not ideally steady, its ripple will superpose on the tested machine, so the measured torque ripple is a little higher than that by FEM. However, the torque ripple of the prototype machine is satisfactory and shows a good design.

5.3. Cogging Torque Test

In this paper, the method in [56] is used to measure the cogging torque. The measurement principle is shown in Figure 14. The balance weight is added to ensure that the metal bar acts on the digital gauge and the displayed values of the digital gauge are always positive. The cogging torque can be reflected in the value of the digital gauge when the stator rotates to different positions. Finally, based on the characteristic of zero average torque of the cogging torque, the cogging torque can be obtained by subtracting the average value of the measured data from every original measured data point. Moreover, the problem of the friction torque can be avoided by this way.

In order to obtain correct and precise results, there are several points to be noted: (1) both metal bars must be at the same height to the horizontal stand; (2) the precision of the digital gauge can reach 0.01 g or above to meet the requirement of cogging torque; (3) the stator can rotate a small angle every step; (4) the stator rotates in the same direction in a test process. The platform for measuring the cogging torque is shown in Figure 15.

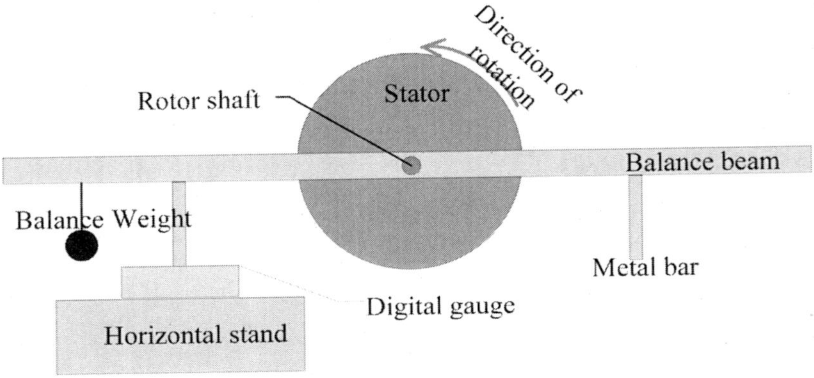

Figure 14. The principle of measurement.

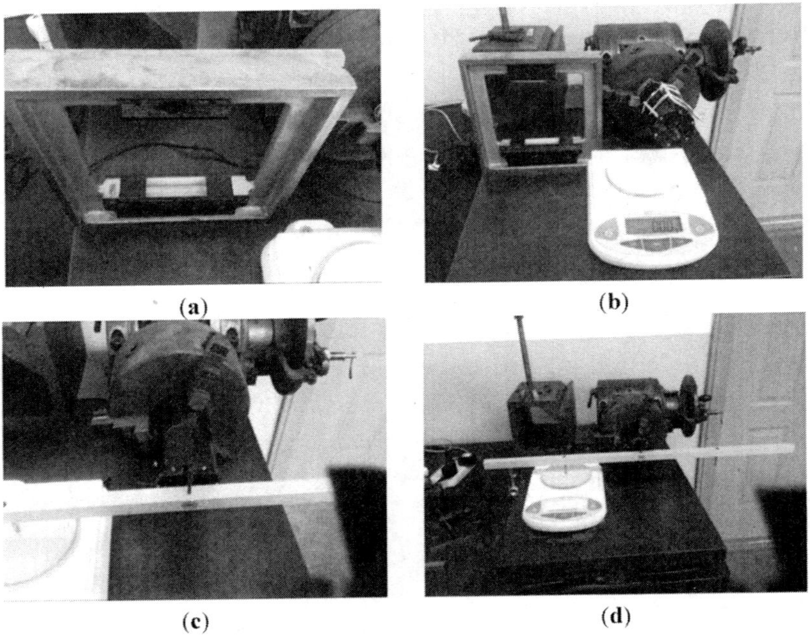

Figure 15. The platform for measuring the cogging torque: (**a**) Frame level; (**b**) Lathe and digital gauge; (**c**) Balance beam; (**d**) All devices.

As is known, the period of the cogging torque waveform is determined by the lowest common multiple (LCM) between the number of slots and the number of permanent magnets. In this paper, the prototype machine has four poles and 18 slots, so the LCM is 36. The period of the cogging torque waveform is 10° in mechanical angle. Figure 16 shows that the waveforms of cogging torque obtained by FEM and the actual measurement are in good agreement. The two waveforms have uniform trends and the peak value is very approximate in one period. An asymmetry air gap and low concentricity between stator and rotor may exist in the machine due to manufacturing errors, so the measured waveform fluctuates a little at some positions compared to the FEM waveform. Both the measured and simulated results indicate a small cogging torque of the prototype machine.

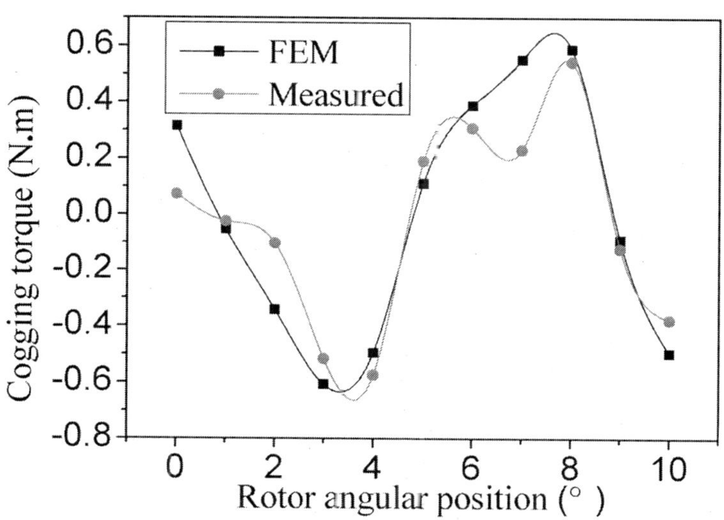

Figure 16. The measured and simulated cogging torque waveforms.

6. CONCLUSIONS

A high speed PMSM for use in a FES system is researched in this paper. By comparing with IPM rotor structure, the rotor structure with SPM is selected for the PMSM. Careful selections of slot/pole matching and magnetic-pole embrace are carried out. Slot/pole matching of 4p-18s and magnetic-pole embrace of 0.81 for N and S poles are obtained considering the torque characteristics. Furthermore, the torque characteristics and back EMF are optimized with Taguchi method on the aspects of magnetic-pole eccentricity, the slot opening, the thickness of PM and the length of air gap. The best combination of D_{off} = 4 mm, b_{s0} = 1.5 mm, T_{PM} = 3 mm, δ = 2.3 mm is obtained after elaborate analysis. After optimization, the THD of the back EMF waveform is decreased by 50.3%, the cogging torque and torque ripple are decreased by 95.1% and 90.9%, respectively, and the average torque increased by 17.9%. A high speed PMSM prototype was manufactured. The back EMFs, torque characteristic at load and cogging torque are verified by experimental results. It is validated that Taguchi method is a good means to optimize the high-speed PMSM. This paper can provide good references for a high speed machine designer during practical design.

ACKNOWLEDGMENTS

This work was supported in part by the National Natural Science Foundation of China under Project 51307008 and 51347009, in part by the Key laboratory for Intelligent Control & Decision of Complex Systems of Beijing Institute of Technology, in part by Ph.D. Programs Foundation of Ministry of Education of China under Project 20121101120024, in part by Basic Research Foundation of Beijing Institute of Technology under Grant

20110642015, 20120642013 and 20130642015, and in part by Excellent Young Scholars Research Fund of Beijing Institute of Technology.

AUTHOR CONTRIBUTIONS

All authors contributed to this work by collaboration. Jing Zhao is the main author of this manuscript. Zhongxin Gu and Xiaobei Li assisted in the design of optimization methods to this study. Bin Li and Zhen Chen provided some useful suggestions in the construction of paper. The whole project was supervised by Xiangdong Liu. All authors revised and approved the publication.

CONFLICTS OF INTEREST

The authors declare no conflict of interest.

REFERENCES

1 Roberts, B.; McDowall, J. Commercial successes in power storage. *IEEE Power Energy Mag.* **2005**, *3*, 24–30.

2 Flynn, M.M.; McMullen, P.; Soliks, O. Saving energy using flywheels. *IEEE Ind. Appl. Mag.* **2008**, *14*, 69–76.

3 Gelman, V. Energy storage that may be too good be true: Comparison between wayside storage and reversible thyristor controlled rectifiers for heavy rail. *IEEE Veh. Technol. Mag.* **2013**, *8*, 70–80.

4 Suvire, G.O.; Mercado, P.E. Combined control of a distribution static synchronous compensator/flywheel energy storage system for wind energy applications. *IET Gener. Transm. Distrib.* **2012**, *6*, 483–492.

5 Schainker, R.B. Executive overview: Energy storage options for a sustainable energy future. In Proceedings of the 2004 IEEE Power Engineering Society General Meeting, Denver, CO, USA, 6–10 June 2004; pp. 2309–2314.

6 Yu, Y.; Wang, Y.; Sun, F. The latest development of the motor/generator for the flywheel energy storage system. In Proceedings of the International Conference Mechatronic Science Electric Engineering and Computer (MEC), Jilin, China, 19–22 August 2011; pp. 1228–1232.

7 McDowall, J.A. Status and outlook of the energy storage market. In Proceedings of the 2007 IEEE Power Engineering Society General Meeting, Tampa, FL, USA, 24–28 June 2007; pp. 1–3.

8 Yubisui, Y.; Kobayashi, S.; Amano, R.; Sugiura, T. Effects of nonlinearity of magnetic force on passing through a critical speed of a rotor with a superconducting bearing. *IEEE Trans. Appl. Supercond.* **2013**, *23*.

9 Subkhan, M.; Komori, M. New concept for flywheel energy storage system using SMB and PMB. *IEEE Trans. Appl. Supercond.* **2011**, *21*, 1485–1488.

10 Diaz-Gonzalez, F.; Bianchi, F.D.; Sumper, A.; Gomis-Bellmunt, O. Control of a flywheel energy storage system for power smoothing in wind power plants. *IEEE Trans. Energy Convers.* **2014**, *29*, 204–214.

11 Roberts, B.P.; Sandberg, C. The role of energy storage in development of smart grids. *Proc. IEEE* **2011**, *99*, 1139–1144.

12 Carrasco, J.M.; Franquelo, L.G.; Bialasiewicz, J.T.; Galvan, E.; Guisado, R.C.P.; Prats, Ma.A.M.; Leon, J.I.; Moreno-Alfonso, N. Power-electronic systems for the grid integration of renewable energy sources: A survey. *IEEE Trans. Ind. Electron.* **2006**, *53*, 1002–1016.

13 Rajapakshe, A.; Madawala, U.K.; Muthumani, D. A model for a fly-wheel driven by a grid connected switch reluctance machine. In Proceedings of the IEEE International Conference Sustainable Energy Technology, Singapore, 24–27 November 2008; pp. 1025–1030.

14 Jaafar, A.; Akli, C.R.; Sareni, B.; Roboam, X.; Jeunesse, A. Sizing and energy management of a hybrid locomotive based on flywheel and accumulators. *IEEE Trans. Veh. Technol.* **2009**, *58*, 3947–3958.

15 Wang, L.; Collins, E.G., Jr.; Li, H. Optimal design and real-time control for energy management in electric vehicles.*IEEE Trans. Veh. Technol.* **2011**, *60*, 1419–1429.

16 Richardson, M.B. Flywheel energy storage system for traction applications. In Proceedings of International Conference on Power Electronics Machines and Drives, Bath, UK, 16–18 April 2002; pp. 275–279.

17 Kenny, B.H.; Kascak, P.E.; Japsen, R.; Dever, T.; Santiago, W. Control of a high-speed flywheel system for energy storage in space applications. *IEEE Tans. Ind. Appl.* **2005**, *41*, 1029–1038.

18 Christopher, D.A.; Beach, R. Flywheel technology development program for aerospace applications. *IEEE Aerosp. Electron. Syst. Mag.* **1998**, *13*, 9–14.

19 Thompson, R.C.; Beno, J.H.; Pak, T.T. Advanced flywheel technology for space applications. In Proceedings of the 37th

Intersociety Energy Conversion Engineering Conference, Washington, DC, USA, 29–31 July 2002; pp. 153–156.

20 Lawrence, R.G.; Craven, K.L.; Nichols, G.D. Flywheel UPS. *IEEE Ind. Appl. Mag.* **2003**, *9*, 44–50.

21 Gurrero, J.M.; de Vicuna, L.G.; Uceda, J. Uninterruptible power supply systems provide protection. *IEEE Ind. Electron. Mag.* **2007**, *1*, 28–38.

22 Kato, S.; Takaku, T.; Sumitani, H.; Shimada, R. Development of voltage sag compensator and UPS using a flywheel induction motor and an engine generator. *IEE J. Trans. Ind. Appl.* **2007**, *167*, 844–850.

23 Toliyat, H.T.; Talebi, S.; McMullen, P.; Co, H.; Filatov, A. Advanced high-speed flywheel energy storage systems for pulsed power applications. In Proceedings of the 1st IEEE Electric Ship Technologies Symposium 2005, Philadelphia, PA, USA, 25–27 July 2005; pp. 379–386.

24 José, R.; Lai, J.-S.; Fang, Z.P. Multilevel inverters: A survey of topologies, controls, and applications. *IEEE Trans. Magn.* **2002**, *49*, 724–738.

25 Yamazaki, K.; Suzuki, A.; Ohto, M.; Takakura, T. Harmonic loss and torque analysis of high-speed induction motors.*IEEE Trans. Ind. Appl.* **2012**, *48*, 933–941.

26 Ko, K.J.; Jang, S.M.; Choi, J.Y.; Lee, S.H.; Han, S.C.; Lee, Y.B. A core loss calculation based on magnetic field analysis considering the time harmonics of high-speed permanent magnet machine according to driving method. In Proceedings of the 2010 International Conference Electrical Machines Systems, Incheon, Korea, 10–13 October 2010; pp. 1143–1146.

27 Steentjes, S.; von Pringsten, G.; Hombitzer, M.; Hamyer, K. Iron-loss model with consideration of minor loops applied to FE-simulations of electrical machines. *IEEE Trans. Magn.* **2013**, *49*, 3945–3948.

28 Gao, N.S.; Yu, L. Research on loss and electromagnetic heat coupling of high speed permanent magnet synchronous motor. In Proceedings of the 2013 IEEE International Conference on Mechatronics and Automation, Takamatsu, Japan, 4–7 August 2013; pp. 81–86.

29 Dong, J.N.; Huang, Y.K.; Jin, L.; Lin, H.Y.; Yang, H. Thermal optimization of a high-speed permanent magnet motor.*IEEE Trans. Magn.* **2014**, *50*, 7018504.

30 Huang, Y.K.; Zhu, J.G.; Guo, Y.G. Thermal analysis of high-speed SMC motor based on thermal network and 3-D FEA with rotational core loss included. *IEEE Trans. Magn.* **2009**, *45*, 4680–4683.

31 Dong, J.N.; Huang, Y.K.; Jin, L. Electromagnetic and thermal analysis of open-circuit air cooled high-speed permanent machines with gramme ring windings. *IEEE Trans. Magn.* **2014**, *50*, 1–4.

32 Boglietti, A.; Cavagnino, A.; Staton, D.; Shanel, M.; Mueller, M.; Mejuto, C. Evolution and modern approaches for thermal analysis of electrical machines. *IEEE Trans. Ind. Electron.* **2009**, *56*, 871–882.

33 Trigeol, J.F.; Bertin, Y.; Lagonotte, P. Thermal modeling of an induction machine through the association of two numerical approaches. *IEEE Trans. Energy Convers.* **2006**, *21*, 314–323.

34 Huang, Y.K.; Dong, J.N.; Zhu, J.G.; Guo, Y.G. Core loss modeling for permanent-magnet motor based on flux variation locus and

finite-element method. *IEEE Trans. Magn.* **2012**, *48*, 1023–1026.

35 Shen, J.X.; Hao, H.; Jin, M.J.; Yuan, C. Reduction of rotor eddy current loss in high speed PM brushless machines by grooving retaining sleeve. *IEEE Trans. Magn.* **2013**, *49*, 3973–3976.

36 Bai, J.; Liu, Y.; Sui, Y.; Tong, C.; Zhao, Q.; Zhang, J. Investigation of the cooling and thermal-measuring system of a compound-structure permanent-magnet synchronous machine. *Energies* **2014**, *7*, 1393–1426.

37 Seo, J.H.; Chung, T.-K.; Lee, C.G.; Jung, S.-Y.; Jung, H.-K. Harmonic iron loss analysis of electrical machines for high-speed operation considering driving condition. *IEEE Trans. Magn.* **2009**, *45*, 4656–4659.

38 Dorrell, D.G.; Popescu, M. Odd stator slot numbers in brushless DC machines—An aid to cogging torque reduction.*IEEE Trans. Magn.* **2011**, *47*, 3012–3015.

39 Hwang, C.C.; Wu, M.H.; Cheng, S.P. Influence of pole and slot combinations on cogging torque in fractional slot PM motors. *J. Magn. Magn. Mater.* **2006**, *304*, e430–e432.

40 Hsiao, C.-Y.; Yeh, S.-N.; Hwang, J.-C. A novel cogging torque simulation method for permanent-magnet synchronous machines. *Energies* **2011**, *4*, 2166–2179.

41 Dorrell, D.G.; Hsieh, M.F.; Popescu, M.; Evans, L.D.; Staton, A.; Grou, V. A review of the design issues and techniques for radial-flux brushless surface and internal rare-earth permanent magnet motors. *IEEE Trans. Ind. Electron.* **2011**, *58*, 3741–3756.

42 Jang, S.M.; Park, H.I.; Choi, J.Y.; Ko, K.J.; Lee, S.H. Magnet pole shape design of permanent magnet machine for minimization

of torque ripple based on electromagnetic field theory. *IEEE Trans. Magn.* **2011**, *47*, 3586–3589.

43 Hsieh, M.F.; Hsu, Y.S. An investigation on influence of magnet arc shaping upon back electromotive force waveforms for design of permanent-magnet brushless motors. *IEEE Trans. Magn.* **2005**, *41*, 3949–3951.

44 Li, Y.; Xing, J.; Wang, T.; Lu, Y. Programmable design of magnet shape for permanent-magnet synchronous motors with sinusoidal back EMF waveforms. *IEEE Trans. Magn.* **2008**, *44*, 2163–2167.

45 Wang, D.; Wang, X.; Kim, M.K.; Jung, S.Y. Integrated optimization of two design techniques for cogging torque reduction combined with analytical method by a simple gradient descent method. *IEEE Trans. Magn.* **2012**, *48*, 2265–2276.

46 Zheng, P.; Zhao, J.; Han, J.Q.; Wang, J.; Yao, Z.Y.; Liu, R.R. Optimization of the magnetic pole shape of a permanent-magnet synchronous motor. *IEEE Trans. Magn.* **2007**, *43*, 2531–2533.

47 Choi, J.S.; Izui, K.; Nishiwaki, S.; Kawamoto, A.; Nomura, T. Topology optimization of the stator for minimizing cogging torque of IPM motors. *IEEE Trans. Magn.* **2011**, *47*, 3027–3027.

48 Xia, J.K.; Dong, T.; Wang, C.Y.; Zhao, J.Y. Low speed high torque PMSM design based on unequal teeth structure. In Proceedings of the 2008 11th International Conference on Electrical Machines and Systems (ICEMS), Wuhan, China, 17–20 October 2008; pp. 3274–3277.

49 Lukaniszyn, M.; JagieLa, M.; Wrobel, R. Optimization of permanent magnet shape for minimum cogging torque using a genetic algorithm. *IEEE Trans. Magn.* **2004**, *40*, 1228–1231.

50 Hemmati, S.; Shokrikojoori, S.; Ghobadi, R.; Ghiasi, M.I. A practical approach to cogging torque reduction in a permanent magnet synchronous motor using non-dominated sorting genetic algorithm. In Proceedings of the 2013 4th Annual International Power Electronics, Drive Systems & Technologies Conference, Tehran, Iran, 13–14 February 2013; pp. 88–92.

51 Hwang, C.C.; Li, Y.L.; Liu, C.T.; Li, P.L. Optimal design of an SPM motor using genetic algorithms and Taguchi method. *IEEE Trans. Magn.* **2009**, *44*, 4325–4328.

52 Lee, S.J.; Kim, K.S.; Cho, S.; Jang, J.; Lee, T.; Hong, J.-P. Taguchi robust design of back electromotive force considering the manufacturing tolerances in IPMSM. In Proceedings of the 2012 Sixth International Conference on ICEF, Dalian, China, 19–21 June 2012; pp. 1–4.

53 Zhu, L.; Jiang, S.Z.; Zhu, Z.Q.; Chan, C.C. Analytical methods for minimizing cogging torque in permanent-magnet machines. *IEEE Trans. Magn.* **2009**, *45*, 2023–2031.

54 Chang, C.H.; Pal, N.; Lim, W.K.; Lin, J.-J. Comparing several population means: A parametric bootstrap method, and its comparison with usual ANOVA F test as well as ANOM. *Comput. Stat.* **2010**, *25*, 71–95.

55 Zheng, P.; Thelin, P.; Chen, A.; Nordlund, E. Influence of saturation and saliency on the inductance of a four-quadrant transducer prototype machine. *IEEE Trans. Magn.* **2006**, *42*, 1319–1322.

56 Zhu, Z.Q. A simple method for measuring cogging torque in permanent magnet machines. In Proceedings of the 2009 IEEE power & Energy Society General Meeting (PES), Calgary, AB, Canada, 26–30 July 2009; pp. 1–4.

CHAPTER 6

Electromotive force measurement of lanthanides in liquid lead

Takehiro Maruyama[a], Kenji Uda[a], Kimikazu Moritani[a], Takayuki Sasaki[a] & Hirotake Moriyama[b*]

a Department of Nuclear Engineering, Kyoto University, Yoshida-honmachi, Sakyo-ku, Kyoto 606-8501, Japan;

b Research Reactor Institute, Kyoto University, Kumatori, Osaka 590-0494, Japan

ABSTRACT

The activity coefficients of La and Pr dissolved in liquid Pb were determined by the electromotive force (EMF) measurement of the galvanic cell, Ln (solid)|LiCl–KCl–$LnCl_3$|Ln–Pb (solution). The temperature dependence of the activity coefficients of lanthanides is expressed by $\log \gamma = a + b/T$, where a and b are the constants. The excess enthalpy and entropy of lanthanides in the melt were obtained and compared with the literature values. The temperature dependence of the solubility of lanthanides was also obtained from the EMF values of the saturated solution.

KEYWORDS

EMF measurement, lanthanum, praseodymium, liquid lead, activity coefficient, solubility, pyrochemical process

1. INTRODUCTION

Pyrochemical and electrochemical processes using high temperature melts are expected to be much useful for separation and recovery of actinides from fission products, and are now being studied in some countries [1]. An analysis of the excess thermodynamic quantities of lanthanides and actinides in high temperature melts was then made from a systematic point of view for reductive extraction of minor actinides (MAs) [2]. The excess enthalpies and entropies of actinide and lanthanide trichlorides in molten chloride phase were found to be correlated with the ionic radii of metal ions possibly due to complex formation. In the liquid metal phase, on the other hand, the excess enthalpies were explained with the Miedema's atomistic model [3] and the excess entropies were with the vibrational entropy due to alloy formation [4]. Using these correlations and models, some missing values of the excess thermodynamic quantities were evaluated, and the separation factors of MAs from lanthanides were calculated in typical reductive extraction systems of Al, Bi, Cd, and Ga.

For further development of the databases, it is needed to improve both the quantity and quality of the excess thermodynamic data by experimental measurements. The purpose of this study is thus to determine the thermodynamic quantities of lanthanides in liquid lead by using the electromotive force (EMF) measurement method, for which reliable values are considered to be still lacked [2]. The activity coefficients of lanthanides in liquid Pb are calculated from the Nernst equation, and then the excess enthalpies and entropies of lanthanides in liquid Pb are obtained for comparison with the literature values [5,6]. The temperature dependence of the solubility of lanthanides was also obtained from the EMF values of the saturated solution.

2. EXPERIMENTAL

Figure 1 shows the experimental apparatus used in the present study. The EMFs of the galvanic cell of the form, Ln (solid)|LiCl–KCl–$LnCl_3$|Ln–Pb (solution), were measured on varying concentrations of Ln where Ln denotes La or Pr. A pure lanthanide electrode was prepared by welding a tantalum lead to a small rod of lanthanide metal. The alloy electrode was prepared by directly dissolving a small piece of pure lanthanide metal in pure lead. The electrolyte of pure LiCl–KCl eutectic (mole ratio of lithium and potassium = 59/41) was purchased from Anderson Physics Laboratory Engineered Materials Inc., and LiCl–KCl–$LnCl_3$ (3 wt%) was prepared by electrolysis with each pure lanthanide electrode. The tantalum wire (1 mm in diameter) and lanthanide metals (La and Pr) were purchased from Nilaco Corporation, and Pb from Soekawa Chemical Co. Ltd. All other reagents used were of analytical grade purchased from Wako Pure Chemicals Co. Ltd. The experiments were carried out in a glove-box filled with purified Ar where the humidity was kept at <1 ppm.

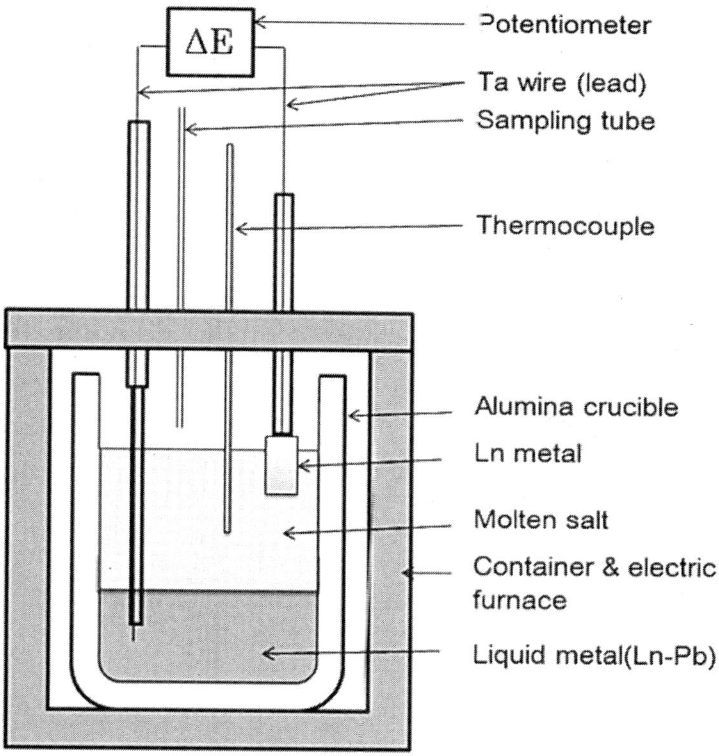

Figure 1. Schematic diagram of the apparatus for EMF measurement.

In the experiment, about 50 g of Pb and a small amount of Ln (La or Pr) were added to an alumina crucible of 34 mm in inner diameter and 150 mm in depth. Also about 50 g of LiCl–KCl was added to it together with a small amount of LiCl–KCl–$LnCl_3$(3 wt%) in order to obtain the solute $LnCl_3$ concentration of 0.3 wt%. The crucible was heated to a given temperature in an electric furnace. After achieving the electrochemical equilibrium (>6 h), the cell EMF was measured with a potentiometer, PARSTAT2263 (Princeton Applied Research). The electrochemical equilibrium was regarded as attained when the EMF changes were less than 0.2 mV within 5 min. After the EMF measurement, a small portion of the metal phase was sucked into a stainless steel tube as a sample. The sample was weighed and dissolved in concentrated HNO_3, and then the Ln concentration was measured by using an inductively coupled plasma mass spectrometry (ICP-MS), ELAN DRC-e (Perkin Elmer). The sample was pretreated for removing Pb in order to reduce the Pb interference during ICP-MS analysis. The concentrations of Ln in the liquid Pb phase were controlled as needed either by electrolysis (for decreasing) or by the addition of metallic Ln (for increasing).

3. RESULTS AND DISCUSSION

3.1. EMF measurements and activity coefficients of lanthanum and praseodymium in liquid lead

The results of the EMF (ΔE) measurements of the galvanic cells, La (solid)|LiCl–KCl–$LaCl_3$|La–Pb (solution) and Pr (solid)|LiCl–KCl–$PrCl_3$|Pr–Pb (solution) are shown in Tables 1 and 2, respectively, together with the evaluated activity coefficients. Standard deviations of the ΔE values are estimated to be around 0.2 mV, and those of the $\log X_{\mathrm{Ln}}$ values are around 0.04. The ΔE values are plotted in Figures 2 and 3, which exhibit typical temperature dependences of the ΔE values of dilute solutions. In fact, the ΔE value of the saturated solution decreases with the increasing temperature and the ΔE value of the unsaturated solution with a constant solute concentration increases with the temperature. As for the measured $\log X_{\mathrm{Ln}}$ values, it is remembered that there are undissolved solid intermetallic compounds in the saturated solutions. In order to avoid any effect of such fine particulate forms of the compounds as dispersed in the solution on the $\log X_{\mathrm{Ln}}$ values, the $\log \gamma_{\mathrm{Ln\ (in\ Pb)}}$ values are obtained from the ΔE and $\log X_{\mathrm{Ln}}$ values of the unsaturated solutions in the present study.

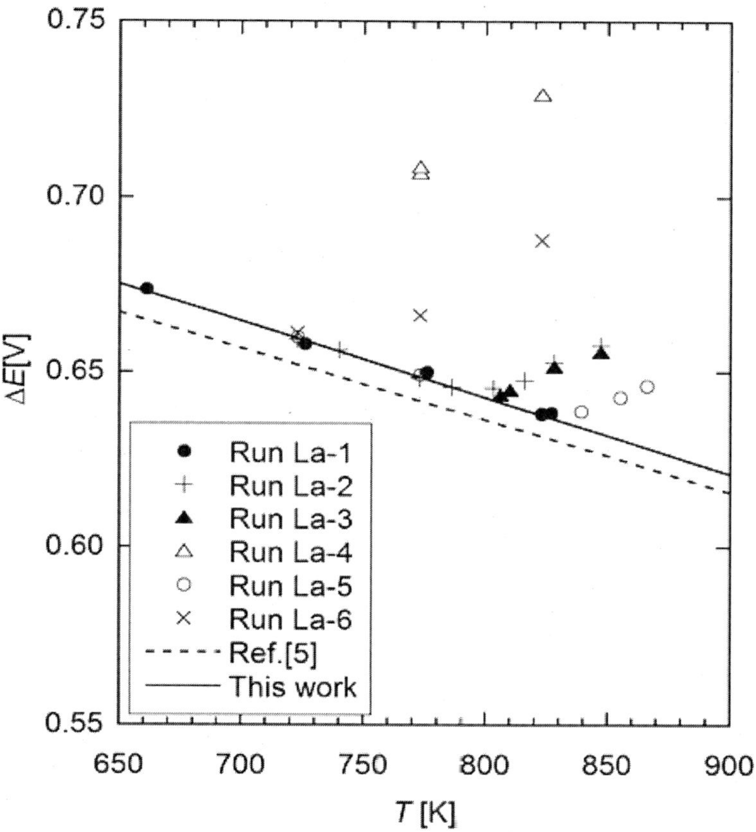

Figure 2. Temperature dependence of the ΔE values of the galvanic cell, La (solid)|LiCl–KCl–$LaCl_3$|La–Pb (solution). Curves represent the ΔE values of saturated La–Pb solution.

The La and Pr ions in chloride melts are in trivalent state, and the electrode reactions of the galvanic cell are represented by:

$$\mathrm{Ln(solid)} = \mathrm{Ln^{3+}(insalt)} + 3\mathrm{e^-} \tag{1}$$

$$\mathrm{Ln}^{3+}(\mathrm{insalt}) + 3\mathrm{e}^- = \mathrm{Ln}(\mathrm{inPb}) \tag{2}$$

According to the Nernst equation and taking into account the Gibbs free energy change of fusion of Ln, the ΔE value is given by:

$$\Delta E = \frac{\Delta G_{\mathrm{fusion}}}{3F} + \frac{2.3RT}{(3F)} \times \log\left(\frac{a_{\mathrm{Ln}}}{a_{\mathrm{Ln\ (in\ Pb)}}}\right) \tag{3}$$

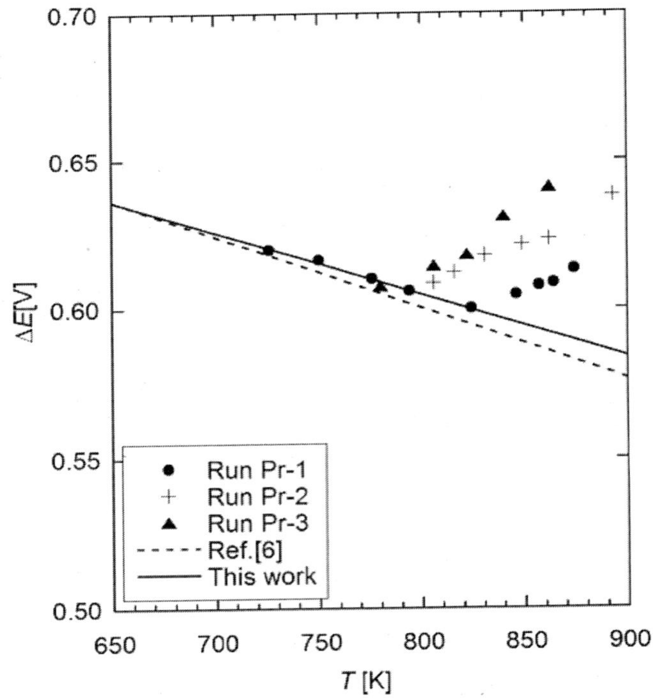

Figure 3. Temperature dependence of the ΔE values of the galvanic cell, Pr (solid)|LiCl–KCl–$PrCl_3$|Pr–Pb (solution). Curves represent the ΔE values of saturated Pr–Pb solution.

Table 1. Results of EMF measurements of the galvanic cell, La (solid)|LiCl–KCl|La–Pb (solution).

Run no.	T [K]	ΔE [V]	ΔG_{fusion} [kJ/mol]*	$\log X_{La\ (in\ Pb)}$	$\log \gamma_{La\ (in\ Pb)}$	Note
La-1	726	0.6580	3.531			Saturated
	823	0.6381	2.814			Saturated
	776	0.6500	3.168			Saturated
	827	0.6385	2.784			Saturated
	661	0.6738	3.976			Saturated
La-2	723	0.6592	3.552			Saturated
	773	0.6485	3.190			Saturated
	828	0.6527	2.776	−3.42	−8.32	
	816	0.6477	2.868	−3.40	−8.42	
	740	0.6562	3.431			Saturated
	786	0.6457	3.094	−3.44	−8.77	
	803	0.6454	2.966	−3.42	−8.54	
	847	0.6576	2.630	−3.42	−8.16	
	773	0.6482	3.190			Saturated
La-3	828	0.6517	2.776	−3.42	−8.30	
	847	0.6558	2.630	−3.43	−8.11	
	806	0.6437	2.944	−3.44	−8.44	
	810	0.6450	2.913	−3.43	−8.42	
La-4	773	0.7071	3.190	−4.69	−8.92	
	773	0.7088	3.190	−4.69	−8.95	
	823	0.7295	2.814	−4.69	−8.53	
La-5	866	0.6462	2.482	−3.22	−7.91	
	723	0.6600	3.552			Saturated
	773	0.6492	3.190			Saturated
	823	0.6383	2.814			Saturated
	839	0.6390	2.692	−3.20	−8.15	
	855	0.6429	2.568	−3.23	−7.98	
La-6	723	0.6611	3.552			Saturated
	773	0.6663	3.190	−4.06	−8.75	
	823	0.6881	2.814	−4.06	−8.40	

Note: *From the literature data [7]: $C_p = 26.44 + 2.33 \times 10^{-3}T$ in J/mol/K, $H_{298} = 0.0$ kJ/mol, and $S_{298} = 56.9$ J/mol/K for solid(hex); $C_p = 17.66 + 15.02 \times 10^{-3}T + 3.89 \times 10^{5}/T^2$ in J/mol/K, $H_{583} = 8.2$ kJ/mol, and $S_{583} = 76.0$ J/mol/K for solid (face-centered cubic, fcc); and $C_p = 34.31$ J/mol/K, $H_{298} = 36.9$ kJ/mol, and $S_{298} = 106.3$ J/mol/K for liquid.

Table 2. Results of EMF measurements of the galvanic cell, Pr (solid)|LiCl–KCl|Pr–Pb (solution).

Run no.	T [K]	ΔE [V]	ΔG_{fusion} [kJ/mol]*	$\log X_{Pr\ (in\ Pb)}$	$\log \gamma_{Pr\ (in\ Pb)}$	Note
Pr-1	794	0.6061	2.815			Saturated
	824	0.6001	2.603			Saturated
	726	0.6200	3.263			Saturated
	750	0.6165	3.111			Saturated
	776	0.6102	2.939			Saturated
	864	0.6086	2.308	−3.56	−6.95	
	874	0.6132	2.233	−3.56	−6.92	
	846	0.6047	2.442	−3.53	−7.12	
	857	0.6077	2.361	−3.60	−6.98	
Pr-2	806	0.6086	2.731	−3.79	−7.45	
	816	0.6122	2.660	−3.78	−7.39	
	831	0.6178	2.552	−3.79	−7.29	
	893	0.6380	2.087	−3.92	−6.76	
	849	0.6215	2.420	−3.83	−7.09	
	862	0.6233	2.323	−3.65	−7.14	
Pr-3	806	0.6142	2.731	−3.76	−7.58	
	822	0.6180	2.617	−3.81	−7.39	
	780	0.6075	2.912			Saturated
	840	0.6310	2.486	−4.08	−7.12	
	862	0.6408	2.323	−4.24	−6.85	

Note: *From the literature data [7]: $C_p = 18.33 + 17.07 \times 10^{-3}T + 3.05 \times 10^{5}/T^2 + 3.97 \times 10^{-6}T^2$ in J/mol/K, $H_{298} = 0.0$ kJ/mol, and $S_{298} = 73.6$ J/mol/K for solid(fcc) and $C_p = 34.31$ J/mol/K, $H_{298} = 40.6$ kJ/mol, and $S_{298} = 127.4$ J/mol/K for liquid.

where ΔE is the measured EMF, ΔG_{fusion} is the Gibbs free energy change of fusion of Ln, and a_{Ln} and $a_{\text{Ln (in Pb)}}$ are the activities of metallic Ln and dissolved Ln in Pb solution, respectively. By adapting $a_{\text{Ln}} = 1$ and $a_{\text{Ln (in Pb)}} = X_{\text{Ln (in Pb)}} \times \gamma_{\text{Ln (in Pb)}}$, Equation (3) is rewritten as:

$$\begin{aligned}\Delta E &= \frac{\Delta G_{\text{fusion}}}{3F} - \frac{2.3RT}{(3F)} \times \log a_{\text{Ln (in Pb)}} \\ &= \frac{\Delta G_{\text{fusion}}}{3F} - \frac{2.3RT}{(3F)} \times \left(\log X_{\text{Ln (in Pb)}} + \log \gamma_{\text{Ln (in Pb)}}\right)\end{aligned} \tag{4}$$

where $X_{\text{Ln (in Pb)}}$ is the mole fraction of Ln in Pb solution, and $\gamma_{\text{Ln (in Pb)}}$ is the activity coefficient of Ln in Pb solution. Equation (4) is the basic thermodynamic expression of this study. The ΔG_{fusion} values are calculated by taking the literature C_p data for fusion of La and Pr [7]. From the measured ΔE and $X_{\text{Ln (in Pb)}}$ values in Tables 1 and 2, the activity coefficient ($\gamma_{\text{Ln (in Pb)}}$) of Ln in Pb solution can be obtained.

Figures 4 and 5 show the temperature dependences of the activity coefficient of La and Pr, respectively. Clear temperature dependence is observed, and the following expression is applied to the observed data:

$$\log \gamma = a + \frac{b}{T} \tag{5}$$

where a and b are experimental constants. Both a and b are determined by the least squares method. Since it is not an ideal mixture of Ln dissolved in Pb solution, there are excess thermodynamic quantities in this system as shown by Equation (8), which agrees with the form of Equation (5).

$$\Delta G_{ex} = 2.3RT \times \log \gamma_{Ln(in\ Pb)} \tag{6}$$

$$\Delta G_{ex} = \Delta H_{ex} - T\Delta S_{ex} \tag{7}$$

$$\log \gamma_{Ln(in\ Pb)} = \frac{\Delta H_{ex} - T\Delta S_{ex}}{2.3RT} = \frac{-\Delta S_{ex}}{2.3R} + \frac{\Delta H_{ex}}{2.3RT} \tag{8}$$

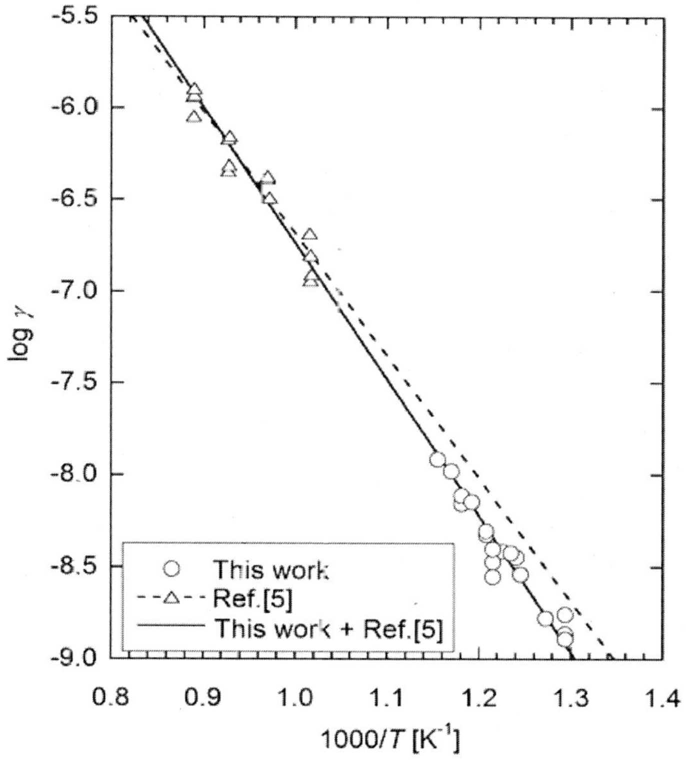

Figure 4. Temperature dependence of activity coefficient of La in liquid Pb.

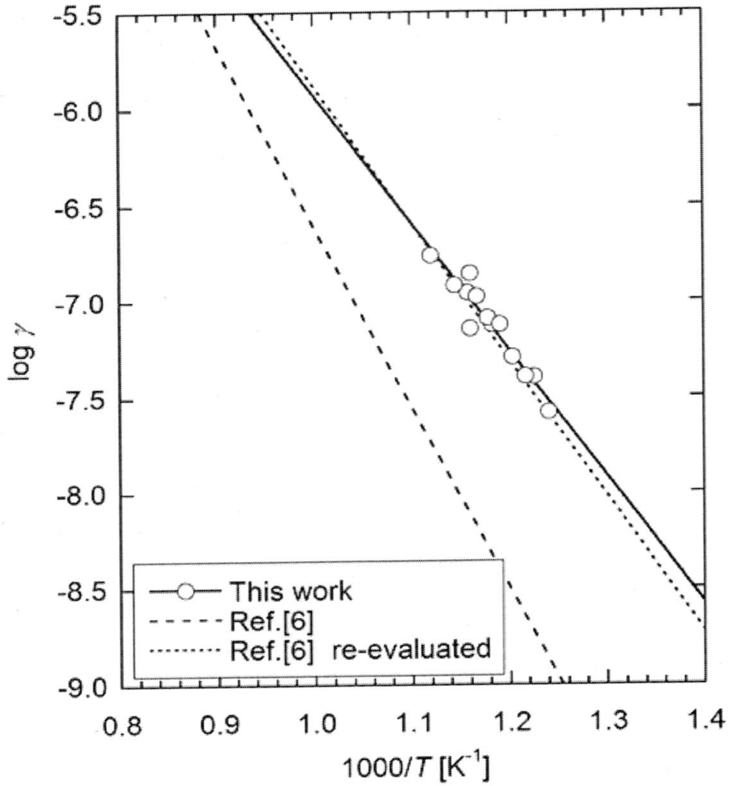

Figure 5. Temperature dependence of activity coefficient of Pr in liquid Pb.

where ΔH_{ex} and ΔS_{ex} are the excess enthalpy and entropy, respectively. By comparing Equation (5) with Equation (8), the ΔH_{ex} and ΔS_{ex} values are obtained by:

$$\Delta H_{ex} = 2.3R \times b \tag{9}$$

$$\Delta S_{ex} = -2.3R \times a \tag{10}$$

By using Equations (9) and (10), the excess enthalpy and entropy of Ln in Pb solution are obtained from the *a* and *b* values determined by the least squares method, as summarized in Table 3.

Table 3. Excess enthalpy and entropy of La and Pr in liquid Pb.

System	Temperature [K]	ΔH_{ex} [M in B] [kJ/mol]	ΔS_{ex} [M in B] [J/mol/K]	Note
La–Pb	723–855	-124.1 ± 8.2	9.1 ± 10.1	This work
	983–1,125	-127.5 ± 9.6	0.5 ± 9.1	Ref. [5]*
	723–1,125	-142.3 ± 1.8	-13.3 ± 2.0	This work + Ref. [5]
Pr–Pb	806–893	-122.7 ± 10.3	-8.6 ± 12.2	This work
	Unknown	-180.5	-54.0	Ref. [6]

Note: *Re-evaluated.

In the case of La in Pb solution, it is noticed that the presently obtained ΔH_{ex} and ΔS_{ex} values are similar to the literature ones [5], but the uncertainties of the values are rather large possibly due to the narrow temperature ranges of the measurements. Then, all the present data and literature ones are used to obtain the ΔH_{ex} and ΔS_{ex} values by the least squares method. Reasonable values are obtained with the improved uncertainties. In order to discuss not only the $\gamma_{Ln\ (in\ Pb)}$ values but also the ΔH_{ex} and ΔS_{ex} values for such a systematic comparison as was made for lanthanides and actinides [2], it is thus important to determine the temperature dependence of the $\gamma_{Ln\ (in\ Pb)}$ values in a broader temperature range and to reduce the measurement errors. In the case of Pr, on the other hand, there is a considerable difference between the presently obtained data and literature ones as shown in Figure 4. For the lack of information on the experimental conditions such as the temperature range in the literature [6], unfortunately, no further discussion is made for this difference. One possible reason may be given below by taking into account the possible overestimation of the literature solubility data.

Figures 6 and 7 show the concentration dependences of the activity coefficient of La and Pr, respectively. Though the number of the data is limited, it is considered that the concentration dependence is only slight and almost negligible in both cases.

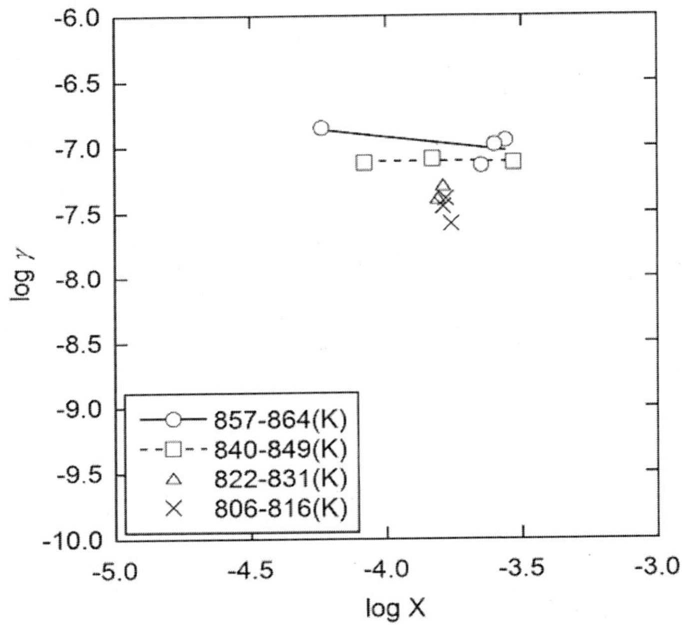

Figure 6. Concentration dependence of activity coefficient of La in liquid Pb.

3.2. Solubility of lanthanum and praseodymium in liquid lead

The solubility of lanthanum and praseodymium in the present liquid lead systems is also obtained. Equation (4) is used to obtain the solubility of these lanthanides from the ΔE values of the saturated solutions by assuming negligible concentration dependence of the activity coefficient as shown in Figures 6 and 7 and by taking the activity coefficient values calculated with the ΔH_{ex} and ΔS_{ex} values in Table 3. In the case of La in Pb solution, the ΔH_{ex} and ΔS_{ex} values with the improved uncertainties are used for the activity coefficient values. The following equations for the solubility are obtained:

$$\log X_{\text{La(in Pb)}} = 2.17 \pm 0.13 - \frac{(4435 \pm 112)}{T} \quad (11)$$

$$\log X_{\text{Pr(in Pb)}} = 2.38 \pm 0.67 - \frac{(4852 \pm 559)}{T} \quad (12)$$

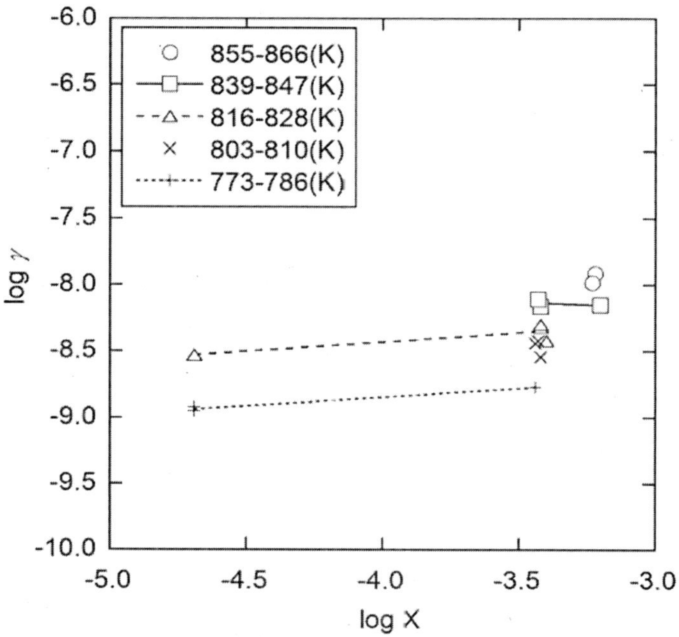

Figure 7. Concentration dependence of activity coefficient of Pr in liquid Pb.

The obtained results are compared with the literature data [5,6]. For La in Pb solution, it is found that Equation (11) gives quite similar values to those calculated by $\log X_{\text{La (in Pb)}} = 2.35 - 4460T$ [5], similarly to the activity coefficients in Figure 4. As for the literature data of Pr in liquid Pb, on the other hand, much higher solubility values are calculated from the equation of $\log X_{\text{Pr (in Pb)}} = 0.81 - 2554/T$ [6]. The literature solubility data might be overestimated possibly due to solid intermetallic compounds in such particulate forms as dispersed in the solution, as mentioned above. It is thus suggested to re-evaluate the $\gamma_{\text{Pr (in Pb)}}$values from the literature EMF data [6] of the saturated solution and the solubility values calculated by Equation (12). As shown in Figure 5, it is found that the re-evaluated $\gamma_{\text{Pr (in Pb)}}$ values are in good agreement with the presently obtained ones. It seems that careful measurements are needed for the solubility values.

4. CONCLUSIONS

The activity coefficients of La and Pr dissolved in liquid Pb were successfully determined by the EMF measurement of the galvanic cell, Ln (solid)|LiCl–KCl–$LnCl_3$|Ln–Pb (solution). From the temperature dependence of the activity coefficients of lanthanides, the excess enthalpy and entropy values of lanthanides in the melt were obtained. It was found that the uncertainties of the values were rather large possibly due to the narrow temperature range of the measurements. The measurements in the broader temperature range were suggested for the determination of the excess enthalpy and entropy values. The solubility of La and Pr in liquid Pb was also determined from the EMF values of saturated solutions by taking the obtained activity coefficient values.

ACKNOWLEDGMENT

This work was partially supported by the Ministry of Education, Culture, Sports, Science and Technology, Grant-in-Aid for Scientific Research (C) 21560873.

REFERENCES

1 International Atomic Energy Agency. 2010. *Assessment of partitioning processes for transmutation of actinides* IAEA-TECDOC-1648, IAEA

2 Toda, T., Maruyama, T., Moritani, K., Moriyama, H. and Hayashi, H. 2009. Thermodynamic properties of lanthanides and actinides in reductive extraction. *J. Nucl. Sci. Technol*, 46: 18–25.

3 de Bore, F. R., Boom, R., Mattens, W. C.M., Miedema, A. R. and Nissen, A. K. 1988. *Cohesion in metals, transition metal alloys*, North-Holland: Amsterdam.

4 Moriyama, H., Kinoshita, K. and Ito, Y. 1990. Equilibrium distribution of actinides and fission products in pyrochemical separation systems, (I) thermodynamic model. *J. Nucl. Sci. Technol*, 27: 827–834.

5 Lebedev, V. A., Kober, V. I., Nichkov, I. F., Raspopin, S. P. and Kalinovsky, A. A. 1972. Study of the thermodynamic properties of La–Al and La–Pb melts by the electromotive force method. *Izv. Akad. Nauk. SSSR, Met*, 2: 91

6 Lebedev, V. A. 1993. *Selectivity of Liquid Metal Electrodes in Molten Halides*, Metallurgiya: Chelyabinsk.

7 Kubaschewski, O., Alcock, C. B. and Spencer, P. J. 1993. *Materials Thermochemistry*, Pergamon Press: New York.

CHAPTER 7

Photo-EMF Sensitivity of Porous Silicon Thin Layer–Crystalline Silicon Heterojunction to Ammonia Adsorption

Yuriy Vashpanov *, Jae Il Jung and Kae Dal Kwack

Electrical and Computer Engineering Division, Hanyang Institute of Technology, Hanyang University, 17 Haengdang-dong, Sung-dong-gu, 133-791 Seoul, Korea

ABSTRACT

A new method of using photo-electromotive force in detecting gas and controlling sensitivity is proposed. Photo-electromotive force on the heterojunction between porous silicon thin layer and crystalline silicon wafer depends on the concentration of ammonia in the measurement chamber. A porous silicon thin layer was formed by electrochemical etching on p-type silicon wafer. A gas and light transparent electrical contact was manufactured to this porous layer. Photo-EMF sensitivity corresponding to ammonia concentration in the range from 10 ppm to 1,000 ppm can be maximized by controlling the intensity of illumination light.

KEYWORDS

gas sensors; porous silicon; heterojunction; photo-EMF

1. INTRODUCTION

Porous silicon technologies have many applications in semiconductor technology, optoelectronics, chemical, biological sensors and other fields of science [1–5]. Electrochemical etching of a silicon wafer surface, under different conditions and additional processing, changes the optical and electrical properties of porous silicon layers widely [6]. Changes in electrical and luminescence properties of the porous silicon under gas adsorptions are well-known [7–12]. However, the influence of the gas adsorptions on photovoltaic junction parameters is not known, yet because of the difficulties of obtaining a sensor with a contact which is transparent to both gases and light. Development of a new method in gas analysis on the base of photovoltaic technology by using the contact can be useful for building new type gas sensors.

Photoluminescence of different visible light spectrum in the porous silicon under UV illumination signalize the appearance of quantum nanowires with band gaps larger than crystalline silicon. This means that a heterojunction is formed between the porous silicon and crystal silicon wafer [13]. Photo electromotive force

(Photo-EMF) can be measured directly by forming a special light transparent contact on the porous surface. This contact is gas transparent too. With use of this gas and light transparent contact, we can study the properties of heterojunction under gas adsorptions by illuminating visible light. In this paper, we report Photo-EMF gas sensitivity of heterojunction between porous silicon and silicon wafer by using the transparent contact.

2. EXPERIMENTAL SECTION

Porous silicon for this study was formed by anode electrochemical etching of (111) oriented p-type silicon wafer with a resistivity of 10 Ωcm at the current density of 10 mA/cm^2, in a HF-based solution [14]. During the etching process an additional illumination and an ultrasonic processing was applied to the silicon surface. The thickness of porous silicon layers is found from cross-section scanning electron microscopy (approximately 12 μm at 5 min etching time). Thin porous film from aluminum was manufactured on a surface of porous layer [gas and light transparent contact on Figure 1(b)]. This film was transparent to illumination and gas molecules and created electrical contact with porous silicon [top electrical contact on Figure 1(b)]. Top to gas and light transparent contact and bottom electrical contact from aluminum to silicon wafer is manufactured by standard technology [top and bottom electrical contacts on Figure 1(b)].

The photo electromotive forces between top and bottom contacts of samples were studied in a special measuring chamber at room temperature [Figure 1(a)]. We used voltmeter-electrometer of B7-30 type for Photo-EMF registration with double screening. Composition of gas atmosphere was changed by gas generator of GR-645 type by dynamical mixing clean nitrogen and ammonia gases [Figure 1 (a)]. The surface of samples was illuminated at different intensity and wavelengths of light. Illumination level was measured by IL Luminance meter T-10 (Konica Minolta). For spectral measurements a spectrophotometer SF26 was used.

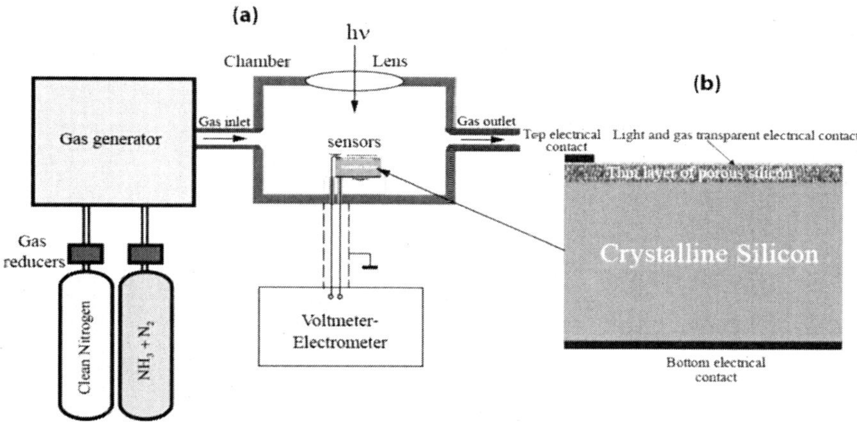

Figure 1. (a) Experimental setup and **(b)** cross-section of Photo-EMF sensors structure.

3. RESULTS AND DISCUSSION

Electrical voltages (Photo-EMF) between top and bottom contacts [Figure 1 (a)] were registered under illumination of samples. Their magnitudes depended on concentration of ammonia in the measurement chamber. Figure 2 shows spectral dependence of Photo-EMF at an illumination level 200 lx at different concentrations of ammonia in the chamber. A maximal magnitude of Photo-EMF was observed near the wavelength of light of approximately 730 nm (Figure 2). An increase of ammonia concentration in the measurement camber leads to a decrease of the Photo-EMF magnitude. Usually photo-detectors have a maximum photo-response at a wavelength of light corresponding to band gap energy E_g [15]. In our case the band gap energy of a porous silicon layer equals approximately 1.7 eV (energy of light quantum at wavelength 730 nm [16]).

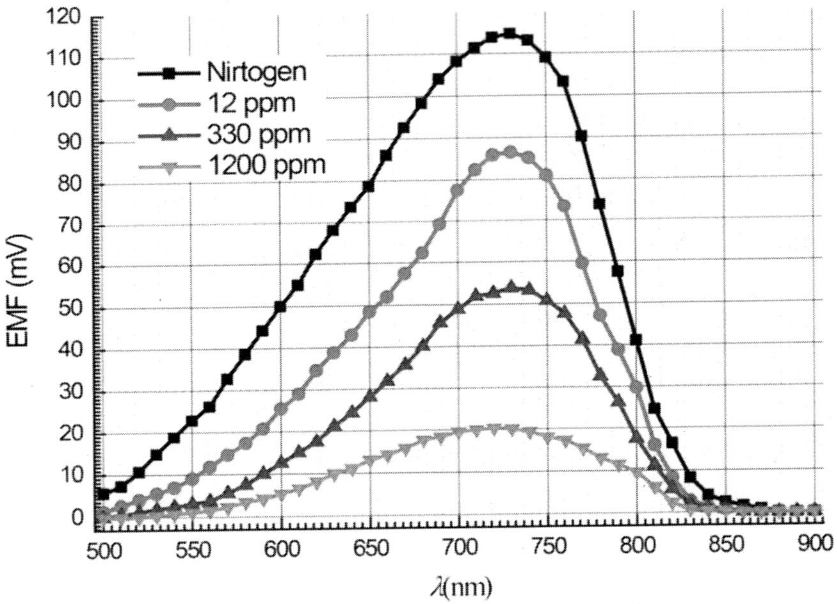

Figure 2. Spectral dependences of Photo-EMF on nitrogen and on different ammonia concentrations.

Manufactured porous silicon layer had red ($\lambda_{max} \approx$ 780 nm) luminescence under laser illuminations at a wavelength of 441.2 nm. This red luminescence had a half-width at half maximum of approximately 0.2 eV. The luminescence in visible range of spectrum is related to quantum wires [1]. Broad band of luminescence peak demonstrates a different thickness of quantum wires. It means that porous silicon have quantum wires with greater band gap than crystalline silicon. Band gaps of our nanowires system are in the ranges of 1.7 ± 0.2 eV. In our case the heterojunction should be formed between porous silicon and silicon wafer. The electrical contacts had ohmic properties. The light induced heterojunction can be the single reason of occurrence of the photo electromotive force on contacts.

Figure 3 shows the experimental dependences on illumination level of the maximal magnitudes of Photo-EMF under different concentrations of ammonia. The magnitude of Photo-EMF increases with growing illumination level (Figure 3). Adsorption of ammonia in porous silicon layer affected the photo-EMF magnitudes appreciably.

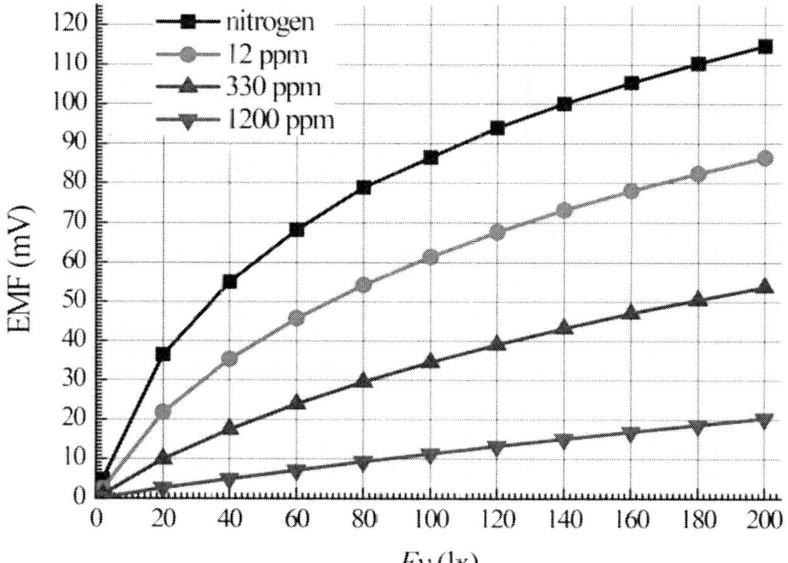

Figure 3. Dependences of maximal magnitudes of Photo-EMF on illumination level under nitrogen and different concentrations of ammonia.

Measurements between top and bottom contacts of current-voltage characteristics showed typical rectifying behavior. The rectification ratio reached a value of 10^5. Under small bias the ideality factor was approximately 2. From the data of current-voltage and spectral dependence of Photo-EMF energy band diagram of heterojunction can be found between crystalline silicon wafer and quantum wire (Figure 4).

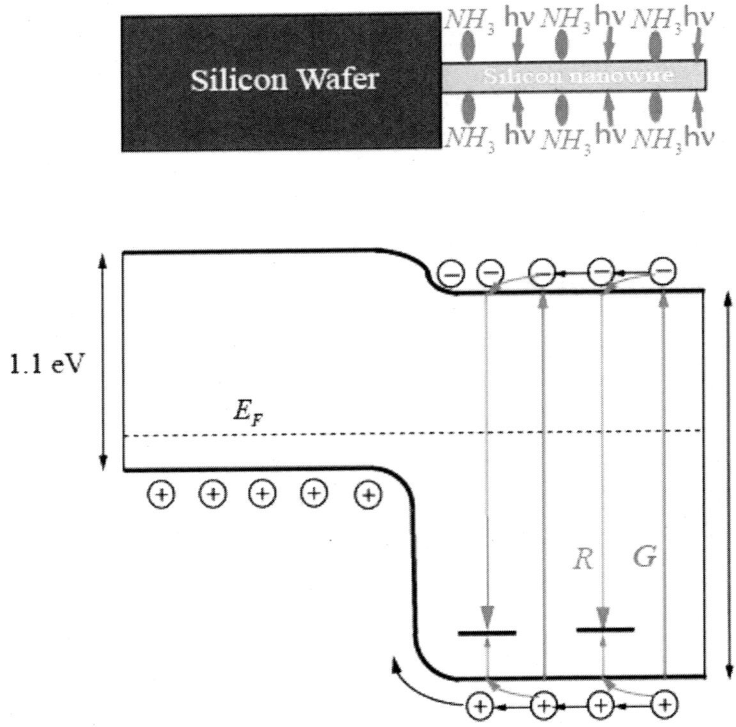

Figure 4. Simplified model of heterojunction energy diagram between silicon nanowire and crystalline silicon wafer under light illumination at wavelength 730 nm and ammonia molecule adsorption.

Illumination of porous silicon layer generate electron-hole pairs G in nanowire (generation process is marked in red in Figure 4). These electron-hole pairs recombined through recombination centers (recombination process is marked in blue in Figure 4). A part of holes reach to p-type region of crystalline silicon. Then photo electromotive force should appear on heterojunction.

The intensity is inhomogeneous along nanowires. The generation of electron-hole pairs is maximal on top part of the porous layer. However, a physical mechanism of Photo-EMF should be identical with well-known photo electromotive force of light induced heterojunction [17].

Ammonia molecules are adsorbed mainly on surface of wires (Figure 4). Adsorption of ammonia molecules creates new surface levels. A re-charging of levels in quantum wires and electrical micro fields close to polar ammonia molecules can affect on recombination rates of electron-hole pairs. In our case, the ammonia adsorption substantially influences on magnitude of photo-EMF (Figures 2 and 3).

Photo electromotive force can be defined as voltage at the heterojunction under illumination in the absence of a current in a circuit. This voltage can be described by the following formula [17]:

$$U = \frac{AkT}{e}\ln\left(\frac{J_{sc}}{J_0}+1\right) \qquad (1)$$

where $J_{sc} = \eta J_{ph}$, J_{ph} is the photocurrent, η is the collection efficiency, J_0 is saturation current in darkness, A is quality factor of heterojunction [18], e is charge of electron, k is Boltzmann constant, T is temperature. In our case quality factor of heterojunction equals approximately 2.

Increase of light intensity leads to increasing voltage U (Photo-EMF) because the photocurrent increases. Decrease of voltage U under adsorption of ammonia may be is related to recombination of electron-hole pairs through recombination levels (collection efficiency η under adsorption decreases).

Figure 5 shows the concentration dependences of Photo-EMF at different level of illumination. Different levels of Photo-EMF are detected for different ranges of concentration: from 100 ppm to 10,000 ppm at 200 lx, from 10 ppm to 1,000 ppm at 20 lx and from 1 ppm to 100 ppm at 2 lx. In the literature on gas sensor technologies (for example, [19]), no information about controlling sensor's sensitivity based on the illumination level control is found.

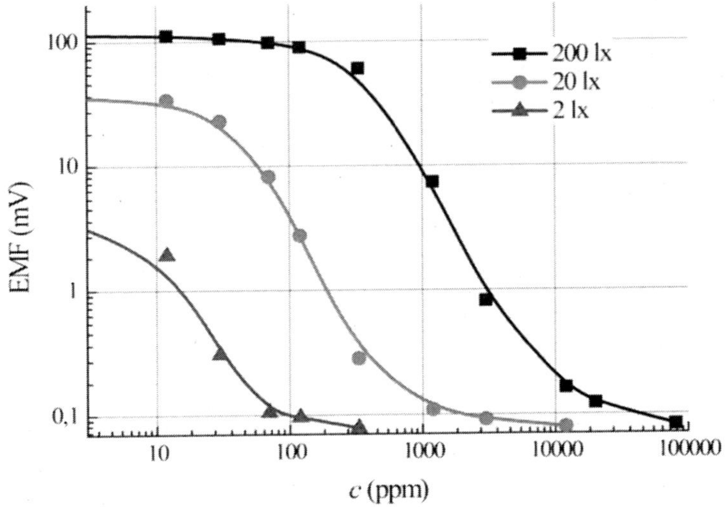

Figure 5. Dependences of Photo-EMF on ammonia concentration under different illumination levels.

Selectivity properties of gas sensors are very important in practice. Porous silicon has an electrical sensitivity to numerous gases. Probably, the photo-EMF induced by ammonia can be observed for other gases too. There are many publications about humidity sensors on the base of porous silicon (see, for example, [20]). Perhaps, sensitivity of porous silicon to humidity could be reduced as compared with ammonia under the operating temperatures more than 100 °C. Illuminating hetero-junction between porous silicon thin layer and crystalline silicon wafer, with a more powerful and higher irradiance laser beam will be an interesting way of studying photo-adsorption-desorption properties of porous silicon surface to different gases. The photo-EMF method of detecting gases could be used in an automatic gas measuring system in chemical industries related with ammonia and nitrogen molecules.

4. CONCLUSIONS

Based on our results, we conclude that photo-electromotive force on the heterojunction between a porous silicon thin layer and a crystalline silicon wafer depends on concentration of ammonia in the measurement chamber. Photo-EMF sensitivity corresponding to ammonia concentration in the range from 10 ppm to 1,000 ppm can be maximized by controlling the intensity of illumination light. The measuring photo-EMF in porous silicon based gas sensor will open a new way of building gas sensors.

ACKNOWLEDGMENTS

This work is supported by Hanyang University's Brain Korea 21 program. Authors are grateful to Jung Young Son for useful discussion and remarks.

REFERENCES

1 Canham, L. *Properties of Porous Silicon*; INSPEC: London, UK, 1997; pp. 1–399.

2 Bisi, O.; Ossicini, S.; Pavesi, L. Porous silicon: A quantum sponge structure for silicon based optoelectronics. *Surf. Sci. Rep* **2000**, *38*, 1–126.

3 Pavesi, L.; Turan, R. *Silicon Nanocrystals. Fundamentals, Synthesis and Applications*; Wiley-VCH: Weinheim, Germany, 2010; pp. 349–394.

4 Stefano, L.D.; Areari, P.; Lamberti, A.; Sanges, C.; Rotiroti, L.; Rea, I.; Rendina, I. DNA optical detection based on porous silicon technology: From biosensors to biochips. *Sensors* **2007**, *7*, 214–221.

5 Kilian, K.A.; Bocking, T.; Lai, L.M.H.; Ilyas, S.; Gaus, K.; Gal, M.; Gooding, J. Organic modification of mesoporous silicon rugate filters: The influence of nanoarchitecture on optical behavior. *Int. J. Nanotechnol* **2008**, *5*, 170–178.

6 Buriak, J.M. *Porous Silicon Technology*; Wiley-VCH: Weinheim, Germany, 2003; pp. 1–400.

7 Ben-Chorin, M.; Kux, A. Adsorbate effects on photoluminescence and electrical conductivity of porous silicon. *Appl. Phys. Lett* **1994**, *64*, 481–483.

8 Baratto, C.; Faglia, G.; Sberveglieri, G.; Gaburro, Z.; Pancheri, L.; Oton, C.; Pavesi, L. Multiparametric porous silicon sensors. *Sensors* **2002**, *2*, 121–126.

9 Galeazzo, E.; Peres, H.E.M.; Santos, G.; Peixoto, N.; Ramirez-Fernandez, F.J. Gas sensitive porous silicon devices: Responses to organic vapors. *Sens. Actuat. B* **2003**, *93*, 384–390.

10 Barillaro, G.; Diligenti, A.; Nannini, A.; Strambini, L.; Comini, E.; Sberveglieri, G. Low-concentration NO_2 detection with an adsorption porous silicon FET. *IEEE Sensors J* **2006**, *6*, 19–23.

11 Li, X.J.; Chen, S.J.; Feng, C.Y. Characterization of silicon nanoporous pillar array as room-temperature capacitive ethanol gas sensor. *Sens. Actuat. B* **2007**, *123*, 461–465.

12 Wang, Y.; Park, S.; Yeow, J.T.W.; Langner, A.; Müller, F. A capacitive humidity sensor based on ordered macroporous silicon with thin film surface coating. *Sens. Actuat. B* **2010**, *149*, 136–142.

13 Islam, M.N.; Ram, S.K.; Kumar, S. Band edge discontinuities and carrier transport across c-Si/Porous silicon heterojunctions. *J. Phys. D Appl. Phys* **2007**, *40*, 5840.

14 Vashpanov, Y.A.; Smyntyna, V.A. *Adsorption Sensitivity of Semiconductors*; Astroprint: Odessa, Ukraine, 2005; p. 178.

15 Koshida, N. Photoconductivity of porous silicon. In *Properties of Porous Silicon*; Canham, L., Ed.; INSPEC: London, UK, 1997; pp. 311–314.

16 Driscoll, W.G.; Vaughan, W. *Handbook of Optics*; McGraw Hill: New-York, NY, USA, 1978; p. B-25.

17 Luque, A.; Hegedus, S. *Handbook of Photovoltaic Science and Engineering*; Wiley: Oxford, UK, 2003; p. 93.

18 Böer, K. *Introduction to Space Charge Effects in Semiconductors*; Springer-Verlag: Berlin, Germany, 2010; p. 283.

19 Soloman, S. *Sensors Handbook*, 2ed ed; McGraw-Hill Companies: New York, NY, USA, 2010; pp. 923–927.

20 Di Francia, G.; Castaldo, A.; Massera, E.; Nasti, I.; Quercia, L.; Rea, I. A very sensitive porous silicon based humidity sensor. *Sens. Actuat. B* **2005**, *111–112*, 135–139.

CHAPTER 8

Development of an Axial Flux MEMS BLDC Micromotor with Increased Efficiency and Power Density

Xiaofeng Ding [1], Guanliang Liu [1,†], Min Du [1,†], Hong Guo [1], Hao Qian [1,*] and Christopher Gerada [2]

[1] School of Automation Science and Electrical Engineering, Beihang University, Beijing 100191, China

[2] Department of Electrical and Electronic Engineering, the University of Nottingham, Nottingham NG7 2RD, UK

ABSTRACT

This paper presents a rigorous design and optimization of an axial flux microelectromechanical systems (MEMS) brushless dc (BLDC) micromotor with dual rotor improving both efficiency and power density with an external diameter of only around 10 mm. The stator is made of two layers of windings by MEMS technology. The rotor is developed by film permanent magnets assembled over the rotor yoke. The characteristics of the MEMS micromotor are analyzed and modeled through a 3-D magnetic equivalent circuit (MEC) taking the leakage flux and fringing effect into account. Such a model yields a relatively accurate prediction of the flux in the air gap, back electromotive force (EMF) and electromagnetic torque, whilst being computationally efficient. Based on 3-D MEC model the multi-objective firefly algorithm (MOFA) is developed for the optimal design of this special machine. Both 3-D finite element (FE) simulation and experiments are employed to validate the MEC model and MOFA optimization design

KEYWORDS

axial flux; microelectromechanical system (MEMS); efficiency; power density; magnetic equivalent circuit (MEC); multi-objective firefly algorithm (MOFA)

1. INTRODUCTION

Micromotors are indispensable for driving microelectromechanical systems (MEMS). There are many types of micromotors, such as electrostatic, piezoelectric and electromagnetic ones. Due to the fact that MEMS evolved from microelectronic technology, most of the prior literature is focused on electrostatic micromotors [1]. Recently, electromagnetic micromotors with a larger torque capability than electrostatic and piezoelectric mircomotors have attracted lots of attention.

Electromagnetic motors at the MEMS scale have not been systematically explored. The first electromagnetic systems assembled with permanent magnets were reported in 2006 [2]. Rare earth magnets such as neodymium-iron-boron (NdFeB) are often used in these micromotors to provide for high energy density. This also indicates that some these machines are often heavily thermally loaded. Given the micro dimensions of these devices, they are susceptible to excessive temperature rises, which in turn degrades the performance and may cause irreversible demagnetization of the magnets [3]. The special windings adopted in these machines are also susceptible to temperature, potentially leading to fusing failures.

In real practice, the output torque of a micromotor is always limited by the thermal load. In reference [1], a micromotor was designed with special emphasis on the maximization of its efficiency. The Joule efficiency obtained is about 42%, which is much higher than those reported for electromagnetic MEMS micromotors [4,5]. However, the capability of its output torque is only 3.78 nNm due to its micro dimensions (external diameter is 2.6 mm). For real applications, such as microrobotics and microaircraft to be competitive, micromotors need to simultaneously have a high output torque density and high efficiency.

Hence, the design of micromotors is typically a nonlinear, multi-objective, optimization problem. Different objectives often conflict with each other, and sometimes a truly optimal solution does not exist. Design compromise and approximation are often needed [6]. Metaheuristic algorithms are very powerful in dealing with this kind of optimization problem [7,8,9,10,11]. The firefly algorithm (FA) is one of the Nature-inspired metaheuristic algorithms proposed by Yang at Cambridge University in 2008 [7]. Multiobjective FA (MOFA) was further developed by Xin-She Yang later in 2012 [11]. Such a multiobjective algorithm is powerful in dealing with design problems in electrical machines with a large number of design variables and multiple objectives under complex nonlinear constraints. Therefore, this paper proposes such an algorithm as a tool for optimizing the design of a micromotor first time.

The remainder of this paper is organized as follows: a description of the axial flux MEMS BLDC micromotor is presented in Section 2. The analytical modeling of the micromotor is developed in Section 3. The stator fabrication following almost-standard MEMS technology is introduced in Section 4. The MOFA optimization for the design of the micromotor targeting both power density and efficiency improvement is presented in Section 5. 3D FEA simulation and experimental verification is reported in Section 6. Conclusions are drawn in Section 7.

2. BASIC STRUCTURE AND OPERATION PRINCIPLE

2.1. The Micromotor Structure

Figure 1 shows the basic structure of an axial flux MEMS micromotor in an exploded geometry view. The stator coils are manufactured on a silicon wafer by a deep etching process, which is standard MEMS technology. The permanent magnets (PMs) with ring shape are assembled over the rotor yokes. The two rotors are symmetrically distributed on two sides of the stator, which improves the output torque as well as avoids any one-sided asymmetric magnetic forces [12]. Furthermore, the stator without the iron core benefits from no iron loss. The surface mounted coils can operate at relatively high current density as they are exposed to a good air flow as the rotor rotates.

In this paper the motor prototype has three phases and 18 coils in total will be shown later. It means that there are six coilsper phase connected in series. Nine coils are distributed symmetrically over each side of the silicon wafer. The upper side coils and corresponding back side coils are connected through via-holes. The three phases are connected in Y topology. The distribution of three phases is shown in Figure 2. Phases A, B and C are distributed counterclockwise. Figure 2 also shows the distribution of the PMs. The PMs are axially magnetized simultaneously. Such a magnetization method leads to a good magnetization field distribution. The rotor consists of three pairs of magnetic poles distributed as shown in Figure 2. In order to avoid excessive flux leakage, the pole-arc coefficient is smaller than 100% [13,14,15].

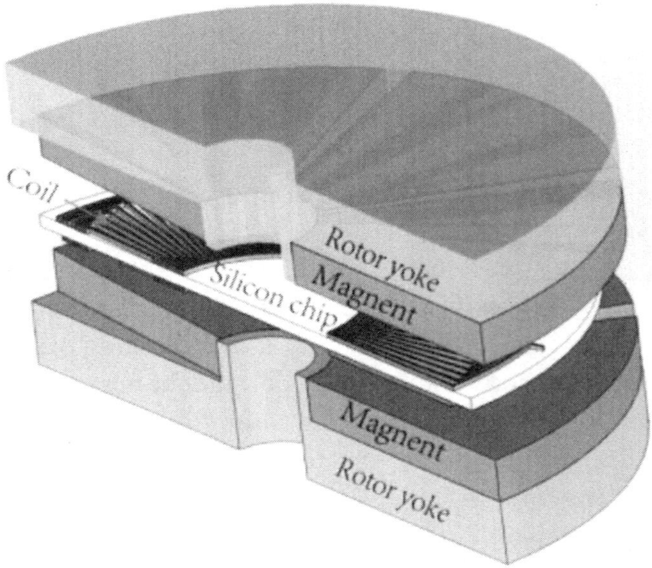

Figure 1. The structure of an axial-flux MEMS micromotor.

2.2. Operation Principle

The operation principle of the micromotor is described in Figure 2. The stator and the rotor on one side of the motor are visible. The other side moves synchronously with the upper side. The magnetizing direction is axial. ⊗ denotes that the magnetic field is orientated in the direction perpendicular to the page pointing inward. ⊙ represents the magnetic field oriented in the direction pointing outwards perpendicular to the page.

Due to the micro dimension of the machine, there is not enough space for assembling any position sensors, such as Hall sensors or resolvers. Therefore, the motor is driven via a senorless approach. During startup, the motor is initially idle and has no back EMF to synchronize the commutator loop. In order to get sufficient back EMF to close the loop, it is necessary to start the motor in open loop. Figure 2a shows the motor startup stage. Currents of phase A and C flow in the clockwise direction, while current of phase B flows in the counterclockwise direction. Then the rotor will be aligned into the position shown in Figure 2a according to the right-hand rule.

At the end of the alignment mode the motor goes into a speed ramping mode. When the back EMF is sufficient the motor will enter the closed-loop running mode. The three-phase voltages are shown in Figure 3. The current flows in the clockwise direction shown in Figure 2 corresponding to the positive voltage shown in Figure 3. When the rotor arrives at the 0° electrical angle as shown in Figure 2b, the current of phase A flows in the clockwise direction while the current of phase C flows in the counterclockwise direction, which depends on the supply voltages of phase A and C at T0 shown in Figure 3. Therefore, the radial conductors of the windings are attracted by the clockwise electromagnetic force due to the magnetic field induced by the PMs of the rotor. In turn, the rotor is attracted by the opposite force, thereby rotating in the counterclockwise direction as shown in Figure 2b. When the rotor reaches a 60° electrical angle as shown in Figure 2c, the current of phase C still flows in the counterclockwise direction. The current of phase A disappears and the current of phase B starts to flow in the clockwise direction. Such currents depend on the voltages of phase B and C at T1 shown in Figure 3. Therefore, the rotor is attracted by a counterclockwise electromagnetic force as well. Such a rotation mechanism is also suitable for the rotor moving to the position shown in Figure 2d. Hence, the rotor rotates continuously in the counterclockwise direction by the corresponding electromagnetic force. The movement of the rotor is demonstrated by a red star ★ shown in Figure 2.

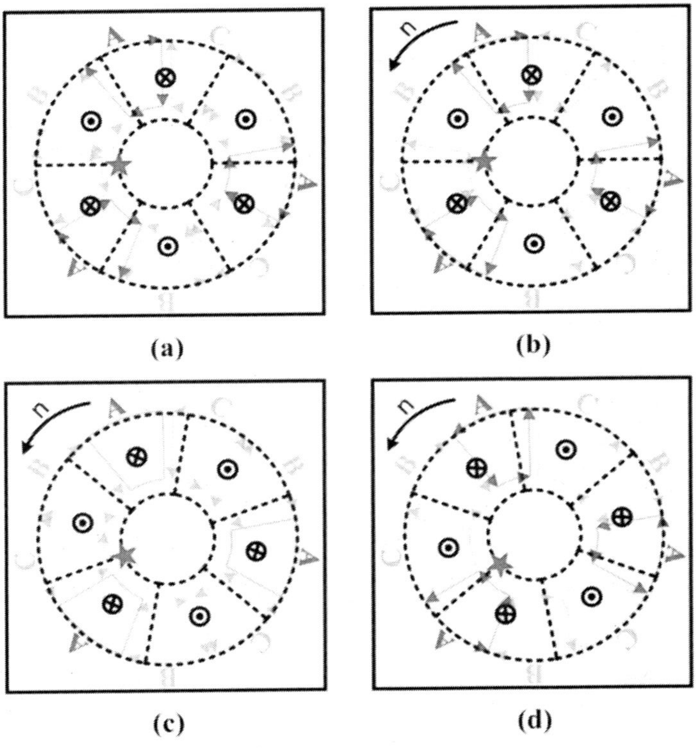

Figure 2. Schematic diagram of the interaction between the stator and rotor windings. (**a**) The initial position; (**b**) 0° electrical angle position; (**c**) 60° electrical angle position; (**d**) 120° electrical angle position.

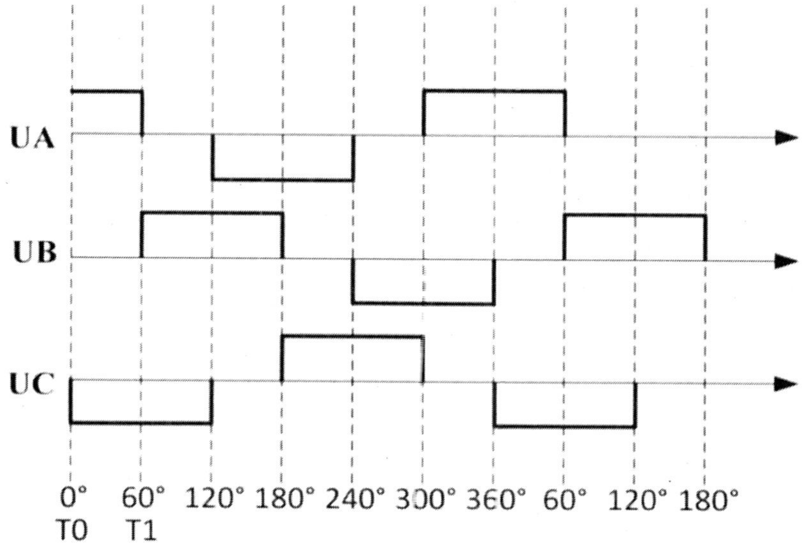

Figure 3. Supply voltage waveforms of the three phases.

3. ANALYTICAL MODELING

A set of equations describing the motor from basic electromagnetic principles are developed. Such equations are used for the optimization design of the micromotor to enhance both the power density and efficiency. In order to simplify the calculation, modeling of the MEMS motor is based on some assumptions as follows:

- There is no saturation occurring in the rotor yokes;
- The magnetic field intensity produced by the armature current in the stator windings is negligible.
- The reluctances of the rotor yokes are neglected.

3.1. Flux Density in Air Gap

The flux density in the air gap generated by the permanent magnets is critical for developing the analytical model. Figure 4a shows the magnetic circuit of the micromotor topology. Magnet-to-rotor and magnet-to-magnet leakage flux is pictured by blue and green lines, respectively, with the main-loop flux in red. A fringing effect exists at the external and internal radius of the magnets, as shown in Figure 4b. The circular-arc straight-line permeance model is used for modeling the flux flow in the air gap [13,14]. The reluctances of magnet to rotorRmr and magnet to magnetRmm are calculated taking the radial length of magnets in 3D space into account. The tangential length of magnets is considered when calculating fringing reluctances Re and Ri. All the above reluctances can be expressed as Equations (1)–(4). The equivalent magnet circuit is proposed in Figure 5.

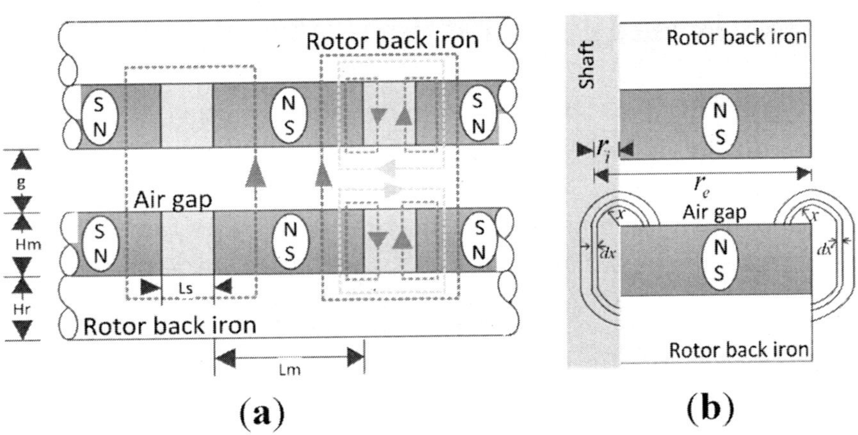

Figure 4. Magnetic loop of the axial-flux micromotor. (**a**) Leakage fluxes; (**b**) Fringing effect.

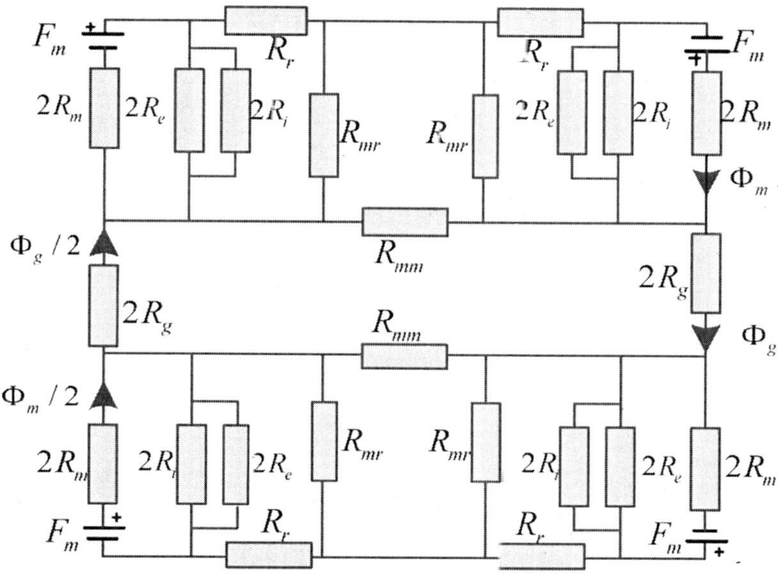

Figure 5. Equivalent magnetic lumped circuit for Figure 4.

$$R_{mr} = 1/[\frac{1}{L_e - L_i}(\int_{L_i}^{L_e} dL_s \int_0^{\frac{L_s}{2}} \frac{\mu_0 L}{H_m + \pi x} dx)] = 1/\{\frac{1}{L_e - L_i}[\int_{L_i}^{L_e} \frac{\mu_0 L}{\pi} \ln(1 + \frac{\pi L_s}{2H_m}) dL_s]\} \tag{1}$$

$$R_{mm} = 1/[\frac{1}{L_e - L_i}(\int_{L_i}^{L_e} dL_s \int_0^{\frac{L_s}{2}} \frac{\mu_0 L}{L_s + \pi x} dx)] = 1/\{\frac{1}{L_e - L_i}[\int_{L_i}^{L_e} \frac{\mu_0 L}{\pi} \ln(1 + \frac{\pi g}{2L_s}) dL_s]\} \tag{2}$$

$$R_e = 1/(\int_0^y \frac{\mu_0 L_{em}}{H_m + \frac{3}{2}\pi x} dx) = 1/[\frac{\mu_c L_{em}}{\frac{3}{2}\pi} \ln\left(1 + \frac{3\pi y}{2H_m}\right)] \tag{3}$$

$$R_i = 1/(\int_0^y \frac{\mu_0 L_{im}}{H_m + \frac{3}{2}\pi x} dx) = 1/[\frac{\mu_0 L_{im}}{\frac{3}{2}\pi} \ln\left(1 + \frac{3\pi y}{2H_m}\right)] \quad (4)$$

$$F_m = H_c H_m = \frac{B_r H_m}{\mu_0 \mu_r} \quad (5)$$

$$R_g = \int_{L_{s_{min}}}^{L_{s_{max}}} \frac{g}{\mu_0 L_s L} dL_s \quad (6)$$

where, Lem=θ·rg, Lim=θ·rg and L=rg−ri, θ is the radian of one magnet, $L_g = (\frac{\pi}{P} - \theta) \cdot r_g$, $L_i = (\frac{\pi}{P} - \theta) \cdot r_i$, and P is the number of pole pairs. The variable *y* depends on the motor dimension and the magnetic material properties. Fm, Rg and Rr represent the equivalent magnetomotive force source, reluctance corresponding to Φg and reluctance of the rotor back iron, respectively. The above equivalent magnetic lumped circuit can be simplified to give the following circuit in the Figure 6:

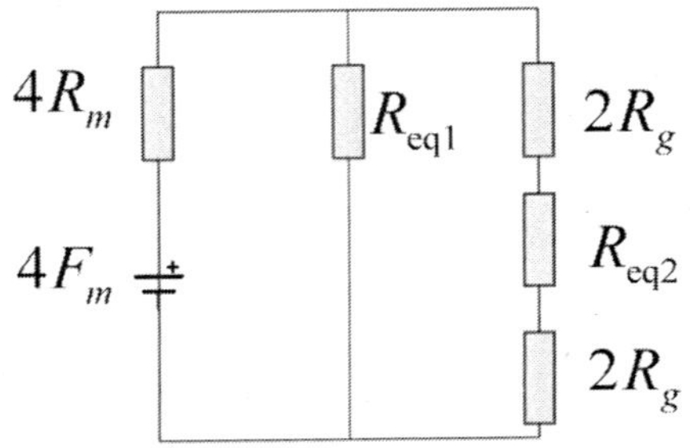

Figure 6. Simplified equivalent magnetic lumped circuit.

Req1 and Req2 represent the equivalent resistances, which are shown as follows:

$$\frac{1}{R_{eq1}} = \frac{1}{4R_i} + \frac{1}{4R_e} + \frac{1}{2R_{mr}} + \frac{1}{R_{mm}} \tag{7}$$

$$\frac{1}{R_{eq1}} = \frac{1}{4R_i} + \frac{1}{4R_e} + \frac{1}{2R_{mr}} + \frac{1}{R_{mm}} + \frac{1}{4R_m} \tag{8}$$

According to the circuit theorem, magnetic flux density of the air gap can be obtained as:

$$\varphi_g = \frac{4R_g}{4R_g + R_{eq2}} \times \frac{R_{eq3}}{R_{eq3} + 4R_m} \times 4F_m \tag{9}$$

$$\frac{1}{R_{eq3}} = \frac{1}{R_{eq1}} + \frac{1}{4R_g + R_{eq2}} \tag{10}$$

$$B_g = \frac{\varphi_g}{S_g} \tag{11}$$

$$S_g = \frac{(L_{em} - 2L_e) \times r_e - (L_{im} - 2L_i) \times r_i}{2} \tag{12}$$

where φ_g represents the flux in the air gap, calculated by the circuit theorem in the formula Equation (9). Bg is the flux density in the air gap, defined as Equation (11). sg is the effective area of the air gap. The effective area is defined as the entire upper surface of the magnet after subtracting the area of the leakage flux. The flux density is assumed as sinusoidally distributed along the circumferential direction in the air gap. The d-axis is defined as 0° electrical angle. The flux density in any electrical angle is:

$$B = B_\delta cos\theta \tag{13}$$

3.2. Back EMF

Due to the small ratio of air-gap over motor radius, it is reasonable to assume that the axial flux density is constant and the radial flux is 0. One coil of the micromotor is shown in Figure 7. The cross-section of one layer is described in Figure 8. The red parts are conductors and the white parts are insulators. The back EMF of one conductor could be expressed as:

$$\varepsilon_i = \int_{r_{1i}}^{r_{2i}} B_\delta cos\theta_i \omega r dr = \frac{1}{2} B_\delta cos\theta_i \omega (r_{2i}^2 - r_{1i}^2) \tag{14}$$

where:

- εi back EMF of conductor i;
- r1i internal radius of turn i;

- r2i external radius of turn i;
- θi the electrical angle of conductor i;
- *i =0,1,2,3…k-1*;
- *k* the number of turns in one coil;
- ω rotation angle velocity.

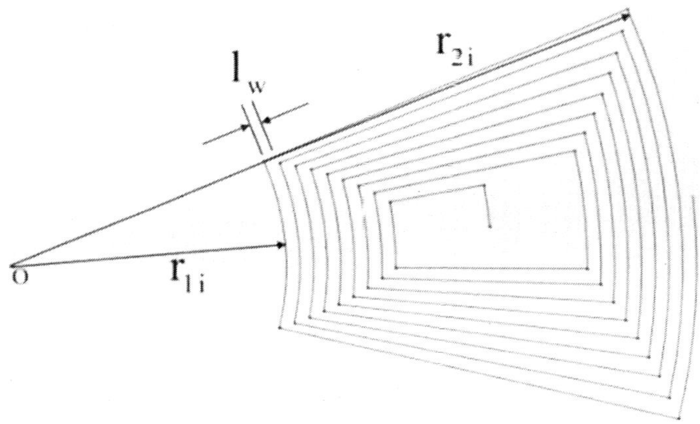

Figure 7. One coil structure of the micromotor.

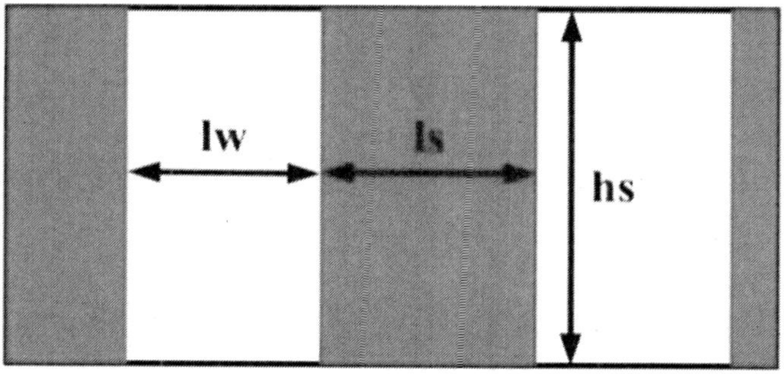

Figure 8. The cross-section of one layer of conductors.

Since there are k turns per coil, the back EMF of one coil is derived as:

$$\varepsilon = \frac{1}{2}\omega B_\delta \sum_{i=0}^{k-1}[\cos\left(-\frac{\theta}{2}+\Delta\theta\cdot i\right)\cdot(r_{2i}^2 - r_{1i}^2) + \cos\left(\frac{\theta}{2}-\Delta\theta\cdot i\right)\cdot(r_{2i}^2 - r_{1i}^2)] \tag{15}$$

where:

- k the number of turns in one coil;
- $\theta = \frac{P\cdot 2\pi}{n}$ radian of one coil;
- P number of pole pairs;
- n number of coils on one side of silicon wafer;
- $\Delta\theta = p\cdot\frac{l_w}{(r_1+r_2)/2}$;
- lw insulator width;
- r1 external radius of coil;
- r2 internal radius of coil.

The motor has three phases and *2n* coils in total, which means that there are *2n/3* coils per phase connected in series. The back EMF of one phase is:

$$E_\theta = \frac{n}{3}\omega B_\delta \sum_{i=0}^{k-1}[\cos\left(-\frac{\theta}{2}+\Delta\theta\cdot i\right)\cdot(r_{2i}^2 - r_{1i}^2) + \cos\left(\frac{\theta}{2}-\Delta\theta\cdot i\right)\cdot(r_{2i}^2 - r_{1i}^2)] \tag{16}$$

3.3. Average Output Torque

The average back EMF of one phase is:

$$\bar{E} = \frac{\int_{\theta_{starting}}^{\theta_{end}} E_\theta d\theta}{2\pi/3} \tag{17}$$

where $\theta_{starting} = \theta_{max} - \frac{p\pi}{n}$, $\theta_{starting} = \theta_{max} + \frac{p\pi}{n}$. The average back EMF reach the maximum value at $\theta = \theta_{max}$, shown as below:

$$\bar{E} = -\frac{n}{2\pi}\omega B_\delta \sum_{i=0}^{k-1}(r_{2i}^2 - r_{1i}^2)|sin\left(\theta_{max} + \frac{p\pi}{n} - \frac{\theta_i}{k}\right) - sin\left(\theta_{max} - \frac{p\pi}{n} + \frac{\theta_i}{k}\right)| \tag{18}$$

Small value of inductances is negligible [9]. Therefore, current of phase is given as below:

$$I = \frac{U - 2\Delta U - 2\bar{E}}{2R} \tag{19}$$

Here U is the power voltage, ΔU is voltage drop across one MOSFET, R is resistance of one phase. The electromagnetic torque could be calculated as:

$$\bar{T} = \frac{2\bar{E}I}{\omega} \tag{20}$$

3.4. Joule Efficiency of Motor

In order to calculate the joule efficiency of the motor, the phase resistance is developed as:

$$R_{ph} = \rho_{cu} \frac{\sum_{i=0}^{k-1} l_i}{l_s \cdot h_s} \tag{21}$$

Here h_s is height of the conductor, l_s is width of the conductor and l_i is length of turn i given as:

$$l_i = 2(r_{2i} - r_{1i}) + r_{1i} \cdot (\theta - i \cdot \Delta\theta) + r_{2i} \cdot (\theta - i \cdot \Delta\theta) \tag{22}$$

Here, $r_{1i} = r_1 + i \cdot l_w, r_{2i} = r_2 - i \cdot l_w$.

Since at any moment there are two phases conducting, the Joule losses of the whole motor can be computed as:

$$P_J = 2P_{J,ph} = 2R_{ph} \cdot I_{ph,rms}^2 \tag{23}$$

Therefore, the Joule efficiency of the motor is:

$$\eta_J = \frac{P_{emc}}{P_{emc} + P_J} \tag{24}$$

4. MODEL VALIDATION BY FEA SIMULATIONS AND EXPERIMENTS

In this section, the analytical model developed previously is verified by both FEA simulation and experiments. Conventional PCB technology is relatively mature. The manufacturing process is relatively simple and the production costs are low. Therefore, in order to carry out experiments to validate the analytical model of the machine, the micromotor is assembled with the stator based on PCB manufacturing technology. The 3D FE model of the motor is developed in MAXWELL software with the PCB stator shown in Figure 9. There are only five turns per coil due to the limitation imposed by the PCB manufacturing technology. The minimal width of conductor is 0.1 mm, while the minimal width of the conductor could be manufactured to the scale of several μm via MEMS. A comparison between PCB and MEMS is described in Section 4. The PCB stator prototype is shown in Figure 10 with parameters of this micromotor listed in Table 1.

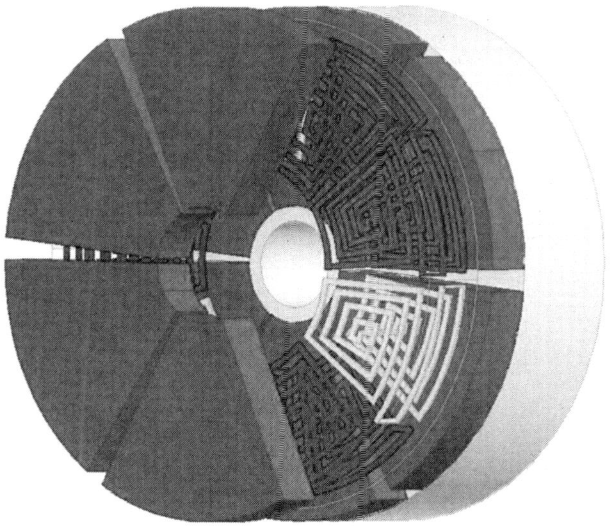

Figure 9. 3D axial flux micromotor with PCB stator.

Figure 10. PCB stator.

Table 1. Specifications of the micromotor.

Parameter	Symbol	Value
Rated current (A)	i	0.1
Rated speed (rpm)	n	20,000
Stator external diameter (mm)	r_e	9.5
Stator internal diameter (mm)	r_i	2.5
Magnet height (mm)	H_m	0.75
Magnet pole arc ratio	-	0.85
Rotor back iron height (mm)	H_r	1.5
Air-gap length (mm)	g	1.6
Pole pairs	p	3

Figure 11 shows the back EMFs obtained from the analytical calculation, FE simulation and experiments, respectively, which show a good match. Other electromagnetic variables are also validated as shown in Table 2. In the worst situation, the error between the analytical model and the experiments is lower than

10%, which also verifies that the analytical model is accurate enough for the motor design optimization.

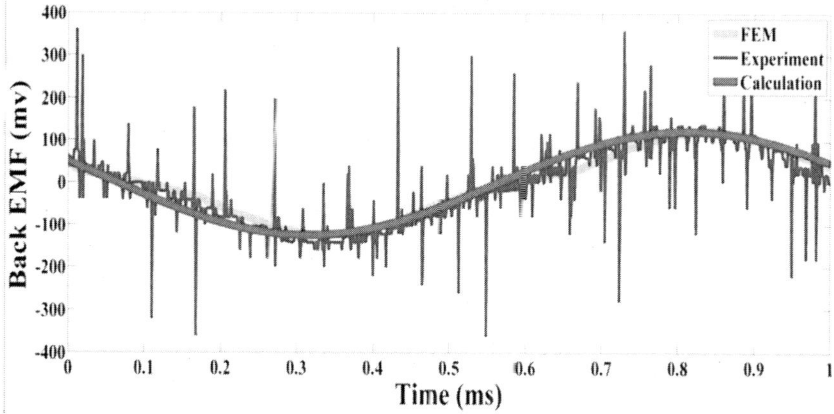

Figure 11. Back EMFs of the micromotor from experiments, FEM and analytical calculations.

Table 2. Comparison between the results obtained with the analytical model and simulations.

Physical Variable	Analytical Model	FEA	Experiment
Flux density, B_δ (T)	0.5797	0.59	-
Torque, T (μNm)	9.68	9.89	10.25
Back EMF, E (mV)	122	128	130.5

5. STATOR FABRICATION

Due to the complex MEMS technology manufacturing process, one stator based on a printed circuit board (PCB) is manufactured and assembled in the prototype micromotor instead of the MEMS stator, which was already presented inSection 3. However, the PCB technology is not suitable for the miniaturization of the micromotor. The minimum width of the copper conductor-based PCB is limited to 0.1 mm. If the width of conductor is 0.1 mm, the maximum height of the conductor is only stacked to 0.1 mm. The ratio of height over width ratio is 1. On the other hand, the minimum width of a copper conductor based on MEMS is less than 10 μm and ratio of height over width ratio is more than 8. The smaller the width of the conductor, the more coils are to be manufactured on the same silicon wafer, which means a higher amplitude of the back EMF and thereby higher output torque. In addition, the higher height over width ratio results in a higher current density of the conductor. Therefore, the output torque is further enhanced by the increment of the current. A comparison between PCB and MEMS technology is shown in Table 3.

Table 3. Comparison between PCB and MEMS.

Parameters	PCB	MEMS
Min width of conductor	0.1 mm	Less than 10 μm
Max height of conductor	0.1 mm	More than 80 μm
Insulator	0.1 mm	Less than 10 μm

Given the advantage of MEMS technology this paper develops a stator based on it. The process flow, which includes more than 20 steps, is quite complicated and time consuming. A simplified version is shown in Figure 12 and described as below.

Step 1: The fabrication starts from a silicon wafer with thickness of 400 μm and a 4 μm silicon oxide (SiO_2) layer. Then spin coating, prebaking, exposure and development of the silicon wafer are implemented in sequence (Figure 12a). The prebaking temperature affects the exposure and development performances directly, so a suitable temperature of 115 °C is selected based on the experimental results;

Step 2: The silicon wafer is baked via a hot plate. The exposed SiO_2 layer is etched off via hydrofluoric (HF) acid (Figure 12b);

Step 3: Grooves with width of 42 μm and depth of 80 μm are etched on the slicon wafer by an ALCATEL601E inductively coupled plasma etcher (ICP) (Figure 12c). Photoresist and SiO_2 are used as a hard mask;

Step 4: A 1 μm aluminum (Al) layer is sputtered over the upper surface of the silicon wafer. Then spin coating, exposure, development and etch are carried out on this surface. The holes are etched in the ICP again after stripping of photoresist (Figure 12d) Al is used to protect the upper surface of the silicon wafer while the down surface is in the fabrication stage. The fabrication process of the down surface is the same as the upper side, therefore only the upper side process is described;

Step 5: A complete silicon groove structure is formed after removing the aluminum cover. After a wet oxidation of the grooves to prevent any short circuits via the silicon substrate, a copper (Cu) seed layer with 200 nm thickness is sputtered over the whole wafer surface (Figure 12e);

Step 6: After a second photolithography and etch, only the cooper seed layer in the grooves is left (Figure 12f);

Step 7: Finally these grooves are filled with the electroplated copper to create the metal lines of the windings (Figure 12g).

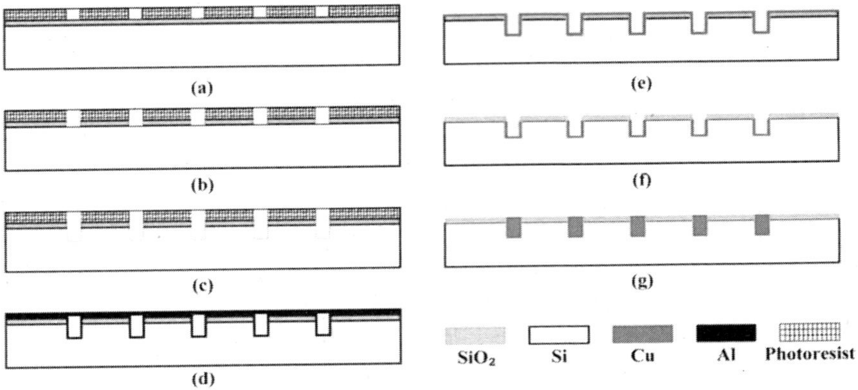

Figure 12. Scheme of the MEMS processing flow. (**a**) silicon wafer with SiO2; (**b**) silicon wafer without portion of SiO2; (**c**) silicon wafer with grooves; (**d**) silicon wafer with Al layer; (**e**) silicon wafer with a wet oxidation of the grooves; (**f**) silicon wafer with the cooper seed layer in the grooves; (**g**) silicon wafer with the metal lines of the windings.

6. FIREFLY ALGORITHM-BASED DESIGN OPTIMIZATION

The FA is one of the Nature-inspired metaheuristic algorithms. The inspiration comes from the flashing light of the firefly. Though the true functions of such signaling systems are still being debated, two fundamental functions of such flashes are to attract mating partners (communication) and potential prey. Based on the flashing characteristics of fireflies the novel firefly algorithm was developed by Xin-She Yang at Cambridge University [7,9,10,11]. Such an algorithm is powerful in dealing with design problems of electrical machines with a large number of design variables and multiple objectives, under complex nonlinear constraints. Therefore, in this paper the MOFA is used to optimize this special micromotor for the first time. Six design variables are shown in Table 4. The geometry constraints and operating limits are shown in Table 5. Current is fixed as 0.1A according to the cooling conditions.

Table 4. Design Variables and Ranges.

Variables	Range	Unit
Stator external diameter	[6~15]	mm
Magnet height	[0.5~2.2]	mm
Number of coils	[6~15]	
Conductor width	[0.04~0.06]	mm
Insulator width	[0.02~0.04]	mm
Power voltage	[1~1.5]	V

Table 5. Design Constraints.

Variables	Value	Unit
Rated current	≤0.1	A
Rated speed	20,000	rpm
Pole pairs	3	
Torque	>35	μNm
Air-gap length (mm)	1.5	mm
Stator internal diameter (mm)	3	mm

The optimization process is described as follows. Firstly, the firefly population is randomly scattered in the solution space. Every firefly emits a distinct brightness light due to its different location. High brightness fireflies can attract low brightness fireflies moving towards them. The distances moved depend on the level of attractiveness. In order to increase the search area and avoid any premature local optimum, the updated location is calculated with an injected disturbance. After moving many times, all individuals will gather at the position of the highest firefly brightness. Hence the optimization results are obtained. The MOFA implementation is based on the following steps:

Step 1: Initialize the numbers of fireflies n, biggest attraction β0, absorption coefficient of light intensity γ, step size factor α, and maximum of iterations or generations tmax.

Step 2: Initialize the positions of fireflies randomly, namely initializing design variables of the machine, the values of objective functions of fireflies are set as their maximum brightness of fluorescence Io.

Step 3: Calculate relative brightness and attractiveness of fireflies, which belong to the population. The direction of movement depends on the relative brightness of fireflies. Here:

$$I = I_0 \times e^{-\gamma r_{ij}} \tag{25}$$

$$\beta = \beta_0 \times e^{-\gamma r_{ij}} \tag{26}$$

where I_0 is the maximum fluorescence brightness of the firefly, namely the fluorescence brightness itself ($r = 0$), which depends on the value of the objective function. β_0 is the maximum attractiveness, namely the attractiveness of the light source ($r = 0$). γ is the absorption coefficient of the light intensity. The fluorescence will gradually weaken according to the increasing distance and the absorption of media. The absorption coefficient of light intensity is set to reflect this feature. rig is the spatial distance between firefly *i* and *j*.

Step 4: Update the spatial positions of fireflies. Random perturbations are injected to the firefly with the best position. The updated equation is:

$$x_i = x_i + \beta \times (x_j - x_i) + \alpha \times (rand - 0.5) \tag{27}$$

where xi, xj represent the spatial positions of firefly *i* and *j*, respectively. α is step size factor. *rand* is random factor distributed uniformly in [0,1].

Step 5: Recalculate the brightness of fireflies according to the updated positions.

Step 6: Return to Step 3 until the search precision is met or the maximum number of generations is achieved.

The proposed optimal design is checked by making a comparison between the reference motor and optimized motor shown in Table 6. Although the dimensions of the machine increase slightly from 367.43 to 439.12 mm3, the improvement of power density T/V is almost 6 Nm/m3. In addition, the increment of the efficiency is more than 8%.

Table 6. Optimization results.

Variables Comparison	Reference	MOFA
Number of coils	9	9
Stator external diameter (mm)	9.6	10.2
PM thickness (mm)	0.8	0.88
Insulator width (mm)	0.04	0.042
Conductor width (mm)	0.04	0.038
Power voltage (V)	1.26	1.5
Performance Comparison	**Reference**	**MOFA**
Dimension (mm^3)	367.43	439.12
Torque ($\mu N \cdot m$)	32.1	40.97
T/V (Nm/m^3)	87.4	93.3
Efficiency	0.53	0.573

7. 3D FEA SIMULATION AND EXPERIMENTAL VERIFICATION

In this section, the optimized design of the machine is verified by both 3D FEA simulation and experiments.

7.1. 3D FEA Simulation Verification

Figure 13 shows the 3D axial flux micromotor with the MEMS stator developed in the MAXWELL software. To demonstrate the improvement of the optimized micromotor, a comparison between the reference and the optimized one is implemented via simulation. The waveforms of the back EMF of the two micromotors are shown in Figure 14. The amplitude of the reference one is 372 mV shown in Figure 14a, while the amplitude of the optimized micromotor reaches up to 450 mV shown in Figure 14b. The increment is more than 77 mV, which is remarkable improvement for the micromotor. Based on the same values of the phase current, the increment of the back EMF represents the increase of electromagnetic torque output. The electromagnetic torque is rising from 32.8 to 41.4 μN·m shown in Figure 15a,b respectively. And the other benefit is given that the torque ripple of the optimized one is smaller than the reference one according to the reasonable distribution of windings. Meanwhile, the efficiency rises from 52.3% to 56.9%. Hence, both efficiency and power density are improved and consistent with the results from MOFA analytical calculation.

Figure 13. 3D axial flux micromotor with MEMS stator.

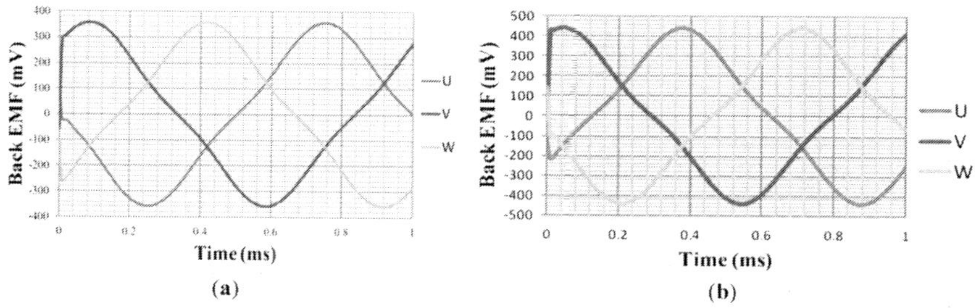

Figure 14. Back EMF waveforms of the two micromotors. (**a**) Back EMF of the reference micromotor; (**b**) Back EMF of the optimized micromotor.

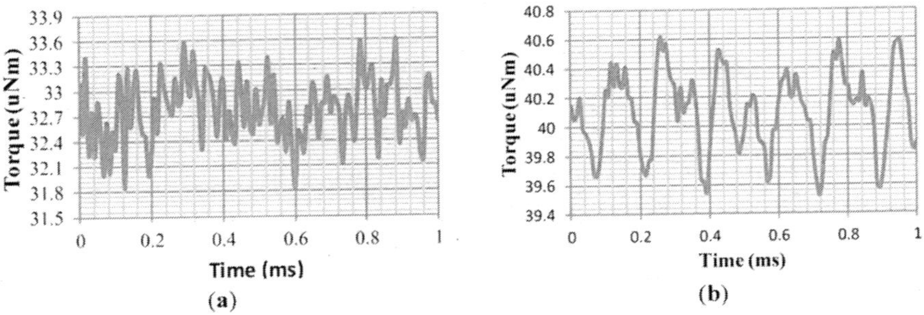

Figure 15. Electromagnetic torques of the two micromotors. (**a**) Electromagnetic torque of the reference micromotor; (**b**) Electromagnetic torque of the optimized micromotor.

7.2. Experimental Verification

A prototype is manufactured according to the optimization design shown in Figure 16a. The stator is fabricated on the silicon wafer in a clean room as described in Section 4. To demonstrate the micro dimensions of the machine, a coin is compared with it as shown in Figure 16b. The diameter of the coin is 19.5 mm. The experimental bench of the MEMS micromotor is shown in Figure 17. The driver is developed based on a sensorless BLDC drive chip ML4426.

Figure 18 shows the PWM control signals for low side drive devices. The PWM duty cycle depends on the output power requirement. The back EMF of phase B is measured as shown in Figure 19. When the speed of the motor is 20,000 rpm the frequency of the back EMF is 1 kHz.

Figure 16. Prototype of MEMS micromotor. (**a**) stator; (**b**) stator with enlarged view; (**c**) dual rotor; (**d**) external view of the micromotor.

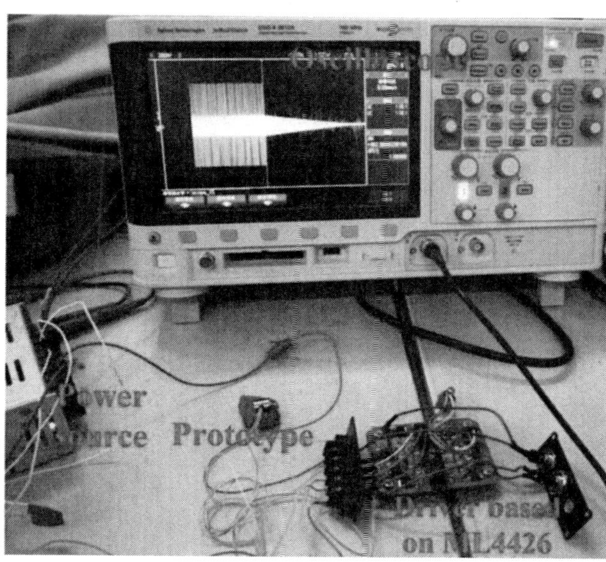

Figure 17. Experimental bench.

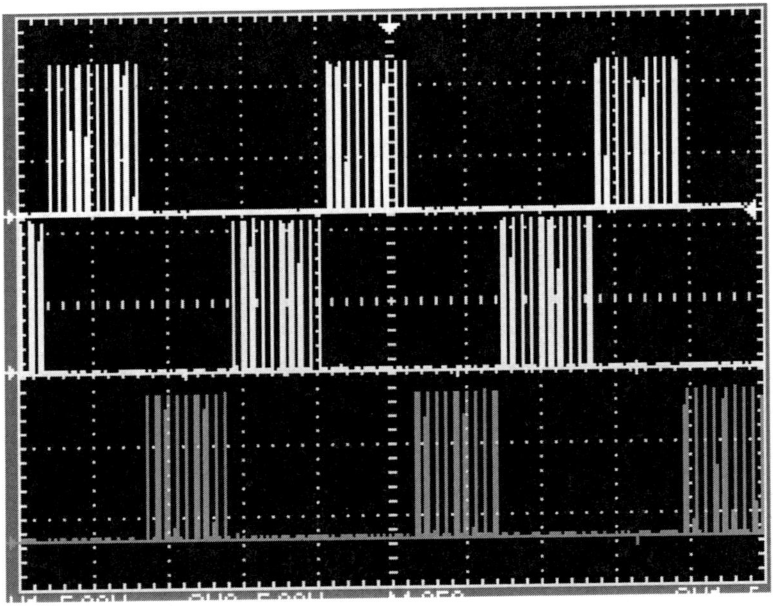

Figure 18. PWM control signals for low side drive devices.

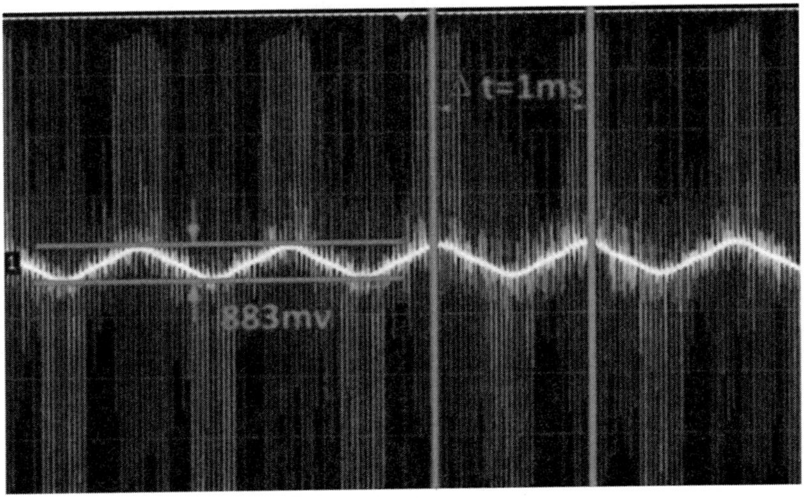

Figure 19. Back EMF of the micromotor @ 20,000 rpm.

The electromagnetic torque is difficult to measure due to the micro dimensions. Therefore a spin-down test is adopted. At a rated speed of 20,000 rpm, the electromagnetic torque is used to overcome the friction and windage torque only. Therefore, the friction and windage torque are equal to the electromagnetic torque. The waveforms of the back EMF during the spin-down of the micromotor are shown in Figure 20. The motor rotational speed is computed as a function of the electric frequency of the back EMF. When the whole system inertia related to the rotor is given, the electromagnetic torque could be calculated as a function of the speed. In this paper, the estimated torque of the micromotor is 40.2 μN·m when the speed of the motor is 20,000 rpm, which is consistent with the simulation result and also validates the analytical calculations used during the MOFA optimization. The resulting efficiency is 56.5% as derived from the experiments. This demonstrates the validity of the modelling approach taken and the MOFA adopted.

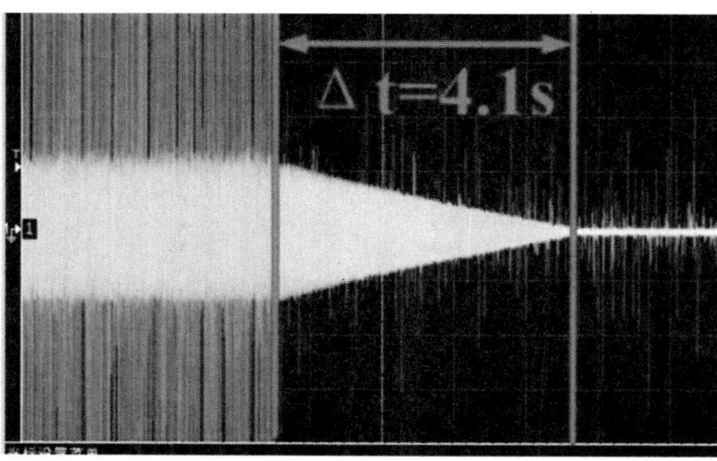

Figure 20. Back EMF measured during the rotor deceleration.

8. CONCLUSIONS

In this paper, an MEMS BLDC micromotor has been developed with improvement of both power density and efficiency. The diameter of the motor is only around 10 mm. Its stator is made of two layers with a high conductor height-over-width ratio, manufactured over a silicon wafer by a MEMS microfabrication process. The rotor is developed by ring shaped PMs assembled over the rotor yoke. A 3D MEC analytical model was developed and used for the optimization process. Such a model proved to be very valuable as it yields an accurate electromagnetic performance prediction of the micromotor as it takes the leakage flux and fringing effect into account. The MOFA is presented for the design optimization of the micromotor. The analytical model and MOFA optimization process for the micromotor are verified by both 3-D FE simulation and experiments. The improvement of power density is reflected in the improvement of torque-to-volume ratio of about 6 Nm/m3 and an increment of efficiency of about 8%. The optimally designed MEMS micromotor shows very good performance characteristics and demonstrates competitive potential for real applications at the MEMS scale.

ACKNOWLEDGMENTS

This work was supported in part by the National Natural Science Foundation of China under Project 51407004 and in part by the Aeronautical Science Foundation of China 2013ZC51031.

AUTHOR CONTRIBUTIONS

Xiaofeng Ding established the 3D MEC analytical model of the MEMS micromotor and performed the simulations with Hao Qian. Guanliang Liu implemented the MOFA optimization design. Min Du did the experiments and stator fabrication. Hong Guo provided some useful suggestions in the construction of the paper framework. The paper was checked by Christopher Gerada. All authors carried out the theoretical analysis and contributed to writing the paper.

CONFLICTS OF INTEREST

The authors declare no conflict of interest.

REFERENCES

1 Merzaghi, S.; Koechli, C.; Perriard, Y. Development of a Hybrid MEMS BLDC Micromotor. *IEEE Trans. Ind. Appl.* **2011**,*47*, 3–11.

2 Achotte, N.; Gilles, P.-A.; Cugat, O.; Delamare, J.; Gaud, P.; Dieppedale, C. Planar brushless magnetic micromotors. *J. Microelectromech. Syst.* **2006**, *15*, 1001–1014.

3 Ding, X.F.; Mi, C. Modeling of Eddy Current Loss and Temperature of the Magnets in Permanent Magnet Machines. *J. Circ. Syst. Comput.* **2011**, *20*, 1287–1301.

4 Wagner, B.; Kreutzer, M.; Benecke, W. Permanent magnet micromotors on silicon substrates. *J. Microelectromech. Syst.* **1993**, *2*, 23–29.

5 Cros, F.; Koser, H.; Allen, M.; Lang, J. Magnetic induction micromachine—Part II: Fabrication and testing. *J. Microelectromech. Syst.* **2006**, *15*, 427–439.

6 Leifsson, L.; Koziel, S. Multi-fidelity design optimization of transonic airfoils using physics-based surrogate modeling and shape-preserving response prediction. *J. Comput. Sci.* **2010**, *1*, 98–106.

7 Yang, X.S. *Nature-Inspired Metaheuristic Algorithms*; Luniver Press: Beckington, UK, 2008.

8 Talbi, E.G. *Metaheuristics: From Design to Implementation*; Wiley: New York, NY, USA, 2009.

9 Yang, X.S. *Engineering Optimisation: An Introduction with Metaheuristic Applications*; Wiley: New York, NY, USA, 2010.

10 Yang, X.S. Firefly algorithms for multimodal optimization. *Stoch. Algorithms Found. Appl.* **2009**, *5792*, 169–178.

11 Yang, X.S. Multiobjective firefly algorithm for continuous optimization. *Eng. Comput.* **2013**, *29*, 175–184.

12 Mahmoudi, A.; Kahourzade, S.; Rahim, N.A.; Hew, W.P. Design, Analysis, and Prototyping of an Axial-Flux Permanent Magnet Motor Based on Genetic Algorithm and Finite-Element Analysis. *IEEE Trans. Magn.* **2013**, *49*, 1479–1492.

13 Ronghai, Q.; Lipo, T.A. Analysis and modeling of air-gap and zigzag leakage fluxes in a surface-mounted permanent-magnet Machine. *IEEE Trans. Ind. Appl.* **2004**, *40*, 121–127.

14 Huang, Y.; Zhou, T.; Dong, J.; Lin, H.; Yang, H.; Cheng, M. Magnetic Equivalent Circuit Modeling of Yokeless Axial Flux Permanent Magnet Machine with Segmented Armature. *IEEE Trans. Magn.* **2014**, *50*, 1–4.

15 Huang, Y.; Ge, B.; Dong, J.; Lin, H.; Zhu, J.; Guo, Y. 3-D Analytical Modeling of No-Load Magnetic Field of Ironless Axial Flux Permanent Magnet Machine. *IEEE Trans. Magn.* **2012**, *48*, 2929–2932.

CHAPTER 9

Influence of Different Rotor Teeth Shapes on the Performance of Flux Switching Permanent Magnet Machines Used for Electric Vehicles

Jing Zhao, Yashuang Yan, Bin Li *, Xiangdong Liu and Zhen Chen

School of Automation, Beijing Institute of Technology, Beijing 100081, China

ABSTRACT

This paper investigated a 12-slot/11-pole flux switching permanent magnet (FSPM) machine used for electric vehicles (EVs). Five novel rotor teeth shapes are proposed and researched to reduce the cogging torque and torque ripple of the FSPM machine. These rotor teeth shapes are notched teeth, stepped teeth, eccentric teeth, combination of notched and stepped teeth, and combination of notched and eccentric teeth. They are applied on the rotor and optimized, respectively. The influences of different rotor teeth shapes on cogging torque, torque ripple and electromagnetic torque are analyzed by the 2-D finite-element method (FEM). Then, the performance of FSPMs with different rotor teeth shapes are compared and evaluated comprehensively from the points of view of cogging torque, torque ripple, electromagnetic torque, flux linkage, back electromotive force (EMF), and so on. The results show that the presented rotor teeth shapes, especially the combination of stepped and notched teeth, can greatly reduce the cogging torque and torque ripple with only slight changes in the average electromagnetic torque.

KEYWORDS

flux switching permanent magnet (FSPM) machine; cogging torque; torque ripple; notched teeth; stepped teeth; eccentric teeth

1. INTRODUCTION

As the number of vehicles increases globally, the energy crisis and environmental pollution have become drastic issues which restrict the development of the world auto industry. Undoubtedly, electric vehicles (EVs) with clean and low energy consumption are one of the best solutions of these problems and have received attention from all around the world. As the key part of EVs, different driveline types have been tried in EVs. Especially, the direct driveline with in-wheel machine is generating more and more interest. It can omit the clutch, the reduction gear and the differential gear. Because in-wheel machines are aimed to provide the traction of EVs directly, they should have high power density, high efficiency and perfect torque characteristic.

The flux switching permanent magnet (FSPM) machine is a potential candidate for the in-wheel machine. It can generate sufficient torque and power density, especially under starting and climbing conditions [1,2,3]. Its high power density characteristics are similar to those of permanent magnet (PM) machines [4,5,6,7,8]. Both windings and PMs are installed on the stator. This structure can avoid both the influence of centrifugal force on PMs and the problem of PM demagnetization caused by heat dissipation issues. As the waveform of the back electromotive force (EMF) is a sine wave, it is preferable to drive the FSPM machine in the brushless alternating current mode. Magnet accumulation effect is generated by neighboring PMs with opposite directions, which can reduce the volume of the machine, lessen the number of the PMs and decrease the machine cost. Different from common PM machines having PMs and armature windings magnetically in series, FSPM machines having PMs and armature windings magnetically in parallel lead to high flux weak capability, which is suitable for a wide speed range [9,10]. In particular, after hybrid excitation flux-switching machines were proposed, the flux linked in the armature windings per phase can be increased or reduced according to the operational status requests, which makes the magnetic field easily adjusted [11,12,13]. The fault-tolerant capability of FSPM machines has also been a hot topic in recent years, FSPM machines with fault-tolerant capability can continue to operate under open circuit and short circuit fault conditions [14,15,16,17]. To sum up, because of the FSPM machine's simple structure, good controllability, high power density and fault-tolerant performance, it is a good selection for driving EVs.

However, the FSPM machine suffers from great slot effects and serious partial saturation due to its own double salient structure, which causes high cogging torque and torque ripple. Numerous methods have been researched for reducing cogging torque and torque ripple considering either the machine design or control aspects. In the machine design aspect, pole and slot matching is a common method to improve the torque performance of machines and it can be directly used in FSPM machines [18,19]. Both finite-element method (FEM) results and experimental tests show that the rotors with odd numbers have lower torque ripple than the rotors with even numbers when the stator structures are the same. Optimization of PM shape is a typical method to improve the torque characteristics of traditional PM machines [20,21]. However, because the PMs are inserted into stator teeth, optimizing the PM shape is not a suitable approach for FSPM machines. Optimizing the rotor structure is however a favorable method for FSPM machines. Rotors with stepped skew were adopted in [22,23] to reduce the cogging torque and torque ripple under various load conditions. Various kinds of notching schemes and their influence on cogging torque are evaluated in [24,25]. It is revealed that the notches effectively reduce torque ripples, but the average torque decreases as well. In order to have a maximum average torque with minimum ripple content, notches with exponential variation profile are imposed [25]. The multi-tooth structure is able to diminish the reluctance of magnetic circuit, which can improve the average torque, reduce the cogging torque and torque ripple [26,27]. The new laminated structure proposed in [28] can decrease magnetic saturation at the edges of stators and rotor teeth and lead to lower cogging torque and higher torque density. In the control aspect, injecting harmonic current or adjusting the current of normal phases can compensate torque ripple [29,30,31], but a sensitive and reliable sensor is needed to confirm the amplitude and angle of the current, and hence the complexity of the entire control system is increased. Optimizing the machine structure can either reduce the cogging torque or reduce the total harmonic distortion (THD) and hence torque ripple, so the machine design method is usually better than the control method.

Considering that the stator structure of a FSPM machine is more complex, a relatively simpler rotor is selected to implement the improvement of torque characteristics of the FSPM machine in this paper. Firstly, the main parameters of FSPM machine are introduced. Secondly, the torque characteristics are analyzed. Under the premise of unchanging the stator's structure, five novel rotor teeth shapes are proposed. Thirdly, the influences of different rotor teeth shapes on cogging torque, torque ripple and electromagnetic torque are analyzed by 2-D FEM using a commercial software package, Ansoft Maxwell. Finally, the performances of FSPMs with different rotor teeth shapes are compared and discussed based on cogging torque, torque ripple, electromagnetic torque, flux linkage, back EMF, and so on. The effects of different rotor teeth shapes are evaluated.

2. MACHINE TOPOLOGIES AND TORQUE ANALYSIS

2.1. Flux Switching Permanent Magnet Machine Main Parameters

A 12-slot/11-pole FSPM machine used for in-wheel EV drive is designed as the initial analysis scheme, as shown in Figure 1.

The main design parameters of the machine are the ratio of the inner diameter to outer diameter of the stator (the split ratio) and the ratio of the rotor pole width to the stator pole width. The former ratio varies in the range of 0.55–0.6 and the latter varies in the range of 1.4–1.6. Detailed specifications are presented in Table 1.

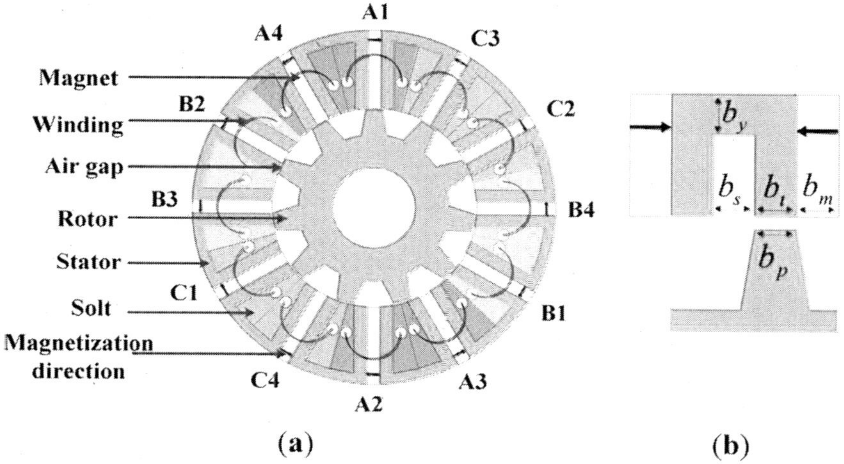

Figure 1. Schematic of the proposed flux switching permanent magnet (FSPM) machine: (**a**) all poles wound cross section; and (**b**) one model.

Table 1. Main design parameters of the FSPM.

Symbol	Machine parameter	Value	Unit
p	Phase number	3	-
N_s	Stator pole number	12	-
N_r	Rotor pole number	11	-
N	Coil number per stator slot	78	-
PM	Magnet material	NTP40UH	-
B_r	Magnet remanence (20 °C)	1.26	T
D_{so}	Stator outer diameter	132	mm
D_{ro}	Rotor outer diameter	72	mm
g	Air gap length	0.3	mm
b_m	Magnet thickness	4.75	mm
h_p	Rotor tooth height	9.5	mm
l_a	Machine stack length	59	mm
n_r	Rated rotational speed	600	rpm
T_{ra}	Rated output torque	15	N·m
I_a	Peak value of rated phase current	15.4	A
U_{dc}	DC link voltage	110	V

2.2. Torque Analysis

The torque curves of the FSPM machine excited with various current amplitudes (3.85–30.8 A) and inner power factor angles (−90°–90°) are shown in Figure 2.

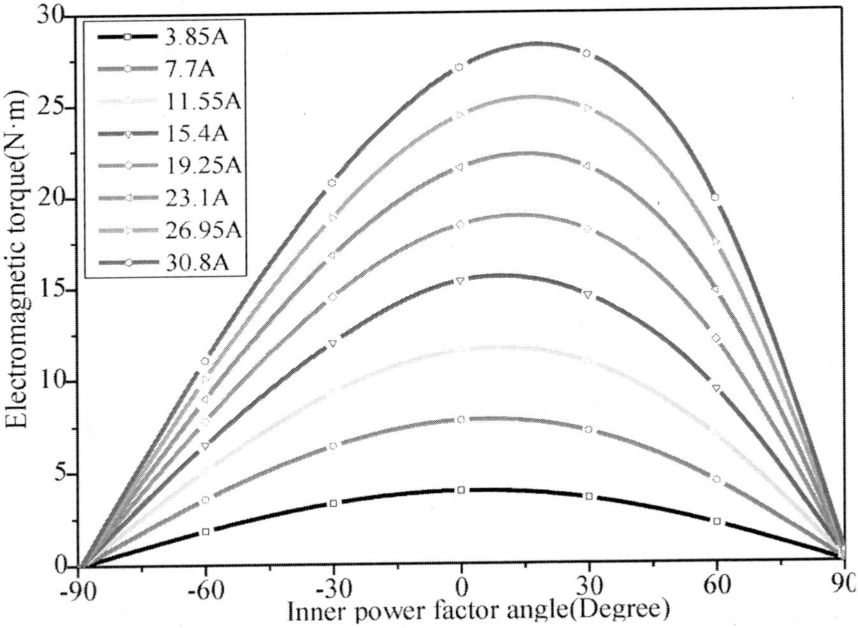

Figure 2. Electromagnetic torques of the FSPM machine with different current amplitudes and inner power factor angles.

The *B-H* curve of the iron core is shown in Figure 3. It can be seen that the torque growth rate is less than the current growth rate due to the saturation of the iron core. The torque reaches its maximum value at around inner power factor angle of 15°.

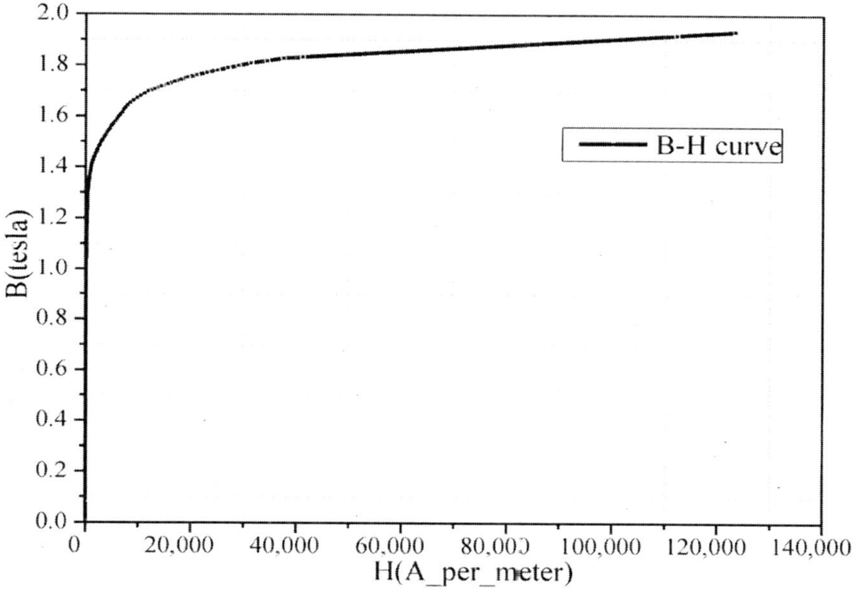

Figure 3. The *B-H* curve of the lamination B35A300.

By the principle of virtual displacement, the electromagnetic torque T_{em} of the FSPM machine can be derived by co-energy [23], the $W_{coenergy}$ can be expressed as:

$$\begin{aligned} W_{coenergy}(i,\theta) &= i^{T}\psi(i,\theta) - W_{field}(i,\theta) \\ &= i^{T}\left[\psi_{PM}(\theta) + L(\theta)i\right] - \left[W_{PM}(\theta) + \frac{1}{2}i^{T}L(\theta)i\right] \\ &= i^{T}\psi_{PM}(\theta) + \frac{1}{2}i^{T}L(\theta)i - W_{PM}(\theta) \end{aligned} \tag{1}$$

where i is the excitation current; θ is the rotor position; $L(\theta)$ is the inductance; $\psi(i,\theta)$ is the flux linkage; W_{field} is the field energy; $\psi_{PM}(\theta)$ is the flux linkage produced by PMs; and W_{PM} is the energy produced by PMs.

Then the electromagnetic torque T_{em} is defined as [23]:

$$\begin{aligned} T_{em} &= \frac{\partial W_{energy}}{\partial \theta} \\ &= i^{T}\frac{d\psi_{PM}(\theta)}{d\theta} + \frac{1}{2}i^{T}\frac{dL(\theta)}{d\theta}i - \frac{dW_{PM}(\theta)}{d\theta} \\ &= T_{PM} + T_{rm} + T_{cog} \end{aligned} \tag{2}$$

where T_{PM} is produced by the interaction of the flux generated by PMs and the armature reaction; T_{rm} is produced by the machine inductance which changes with the rotor position; and T_{cog} is cogging torque generated by PMs and rotor teeth.

It can be seen from Equation (2) that the electromagnetic torque of FSPM machine is composed of three parts: T_{PM}, $T_{rm}$and T_{cog}. However, the cogging torque T_{cog} will not produce effective average torque but only cause torque ripple.

Torque ripple T_r is defined by:

$$T_r = \frac{T_{max} - T_{min}}{T_{avg}} \times 100\% \tag{3}$$

where T_{avg}, T_{max} and T_{min} are average, maximal and minimal torques, respectively.

Therefore, the following section will pay more attention to reducing the cogging torque and torque ripple by modifying the rotor teeth shape to improve the torque performance.

3. THE INFLUENCE OF DIFFERENT ROTOR TEETH SHAPES ON TORQUE PERFORMANCE

In this section, five novel rotor teeth shapes, including notched teeth, stepped teeth, eccentric teeth, the combination of notched and stepped teeth, the combination of notched and eccentric teeth, are applied on the rotor to reduce cogging torque and torque ripple, respectively.

Rotors with notched teeth can lessen the fringing flux prior to the stator and rotor overlap angle. Rotors with stepped teeth can make the waveform of air-gap flux density more close to a sinusoidal wave and the rotors with eccentric teeth are able to weaken the armature reaction. The rotors with a combination teeth shape can have two kinds of effects on cogging torque and torque ripple reduction.

3.1. Rotor with Notched Teeth

Due to the serious partial saturation, a comparatively large torque ripple will be caused by the fringing flux before the stator teeth and the rotor teeth begin overlapping. Figure 4a shows the flux distribution of the initial structure. In order to reduce the fringing flux, this part puts forward a new notched teeth shape. Different from notched teeth at the top of stator or rotor teeth in [16], the proposed notched teeth shape is with v-shape notches on both sides of the rotor teeth. Figure 4b shows the flux distribution of rotor with notched teeth.

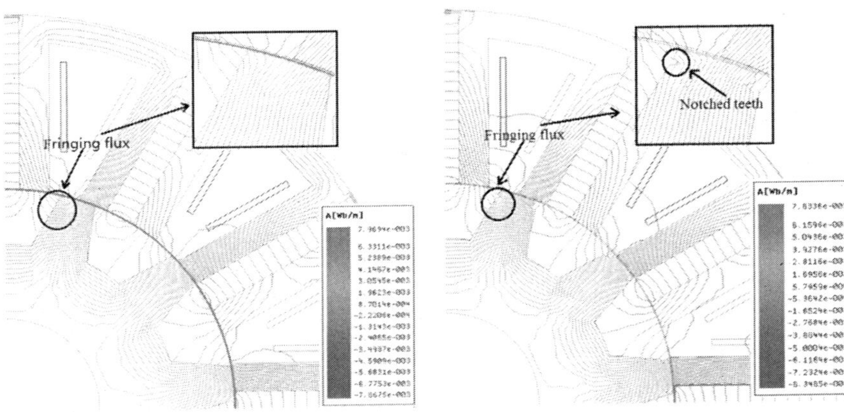

(a) (b)

Figure 4. The flux distribution: (**a**) initial model; and (**b**) rotor with notched teeth model.

Modifications were performed by changing the shape and position of the v-shape notch to obtain better torque characteristics. The ranges of the main dimensions, *i.e.*, the height of the v-shape notch vertex h_1, the depth of v-shape notch$L$, the width of v-shape notch h_2, are shown in Figure 5. Considering that the teeth top has a great influence on the cogging torque, the notch position should not be too far from the teeth top, so the parameter h_1 is selected from 0 mm to 3.0 mm. In addition, as the notches are on the sides of the teeth, the depth and width of the v-shape notch influence the magnetic reluctance directly. Thus, the depth L and width h_2 of the v-shape notch should be small in order to obtain higher average torque. In this paper, L is selected from 0.1 mm to 0.8 mm, and h_2 is selected from 0.04 mm to 0.36 mm.

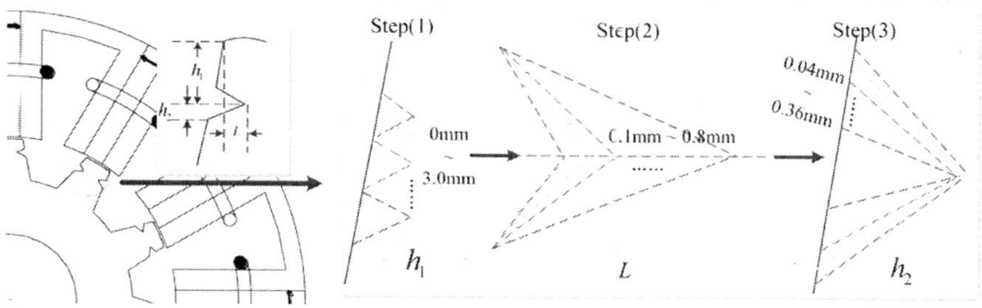

Figure 5. The design procedure of rotor with notched teeth model.

Firstly, changing h_1 at an interval of 0.5 mm from 0 mm to 3.0 mm with h_2 = 0.2 mm and L = 0.5 mm as shown in Step (1) of Figure 5, the waveforms of electromagnetic torques, cogging torques and torque ripples are shown in Figure 6a. It should be noted that the torque data of point h_1 = 0 mm represents the torque data of the initial structure. At the point of h_1 = 1.0 mm, both the cogging torque and torque ripple reach their minimum values with the average torque reduced only by 0.72%.

Secondly, changing L at an interval of 0.1 mm from 0.1 mm to 0.8 mm with h_1 = 1.0 mm and h_2 = 0.2 mm as shown in Step (2) of Figure 5, the torque data *versus* L is shown in Figure 6b. The torque ripple reaches its minimum value at L = 0.5 mm, while the cogging torque is also small.

Finally, changing h_2 at an interval of 0.04 mm from 0.04 mm to 0.36 mm with h_1 = 1.0 mm and L = 0.5 mm as shown in Step (3) of Figure 5, and selecting the final parameters of the v-shape notch from Figure 6c, they are h_1 = 1.0 mm, L = 0.5 mm, andh_2 = 0.28 mm, respectively.

Torque characteristics before and after the notched teeth optimization are compared in Table 2. It can be seen that after improvement by notched teeth, the torque ripple is reduced from 4.09% to 2.56% by 37.41%, the cogging torque is reduced from 187.11 mN·m to 158.97 mN·m by 15.04% and the average electromagnetic torque is reduced by 0.97% compared with the initial structure. The FEM results validate that the notched teeth can reduce the cogging torque and torque ripple obviously with only slight loss of average electromagnetic torque.

3.2. Rotor with Stepped Teeth

The conventional double-salient pole structure leads to an irregular flux density waveform, which results in high cogging torque and torque ripple. Stepped teeth are used to smooth the variation of air-gap length. This method can make the waveform of air-gap flux density more close to a sinusoidal wave, and hence improve the torque characteristics. In order to ensure the symmetry of the rotor teeth, odd portions along the tooth width should be adopted. However, the portions along the tooth width should not be too many considering the complex processing technology. As shown in Figure 7, the step schemes presented in this paper are with 3, 5 and 7 portions along the tooth width, respectively, which is denoted as S1 to S3. They share the same whole width L_S and height H_S. Within one model, each step has equal width and equal height, e.g., $h_{21} = h_{22}$ and $L_{21} = L_{22} = L_{23} = L_{24} = L_{25}$ in S2.

The waveforms of cogging torques, torque ripples and electromagnetic torques are shown in Figure 8. It should be noted that the torque data of point S0 represents the torque data of the initial teeth. It can be seen from Figure 8 that the S2 scheme can reduce the cogging torque and torque ripple, while the S1 scheme and S3 scheme can only reduce torque ripple, but increase the cogging torque, so the S2 scheme is the best.

Torque characteristics before and after the stepped teeth optimization are compared in Table 3. After the S2 scheme is adopted, the cogging torque and torque ripple are reduced to 177.45 mN·m and 3.62% by 5.16% and 11.49% while the average electromagnetic torque is decreased only by 0.25% compared to the initial scheme.

3.3. Rotor with Eccentric Teeth

Eccentric pole, which is usually adopted in PM synchronous machine, is another technique to improve the waveform of air-gap flux density and the torque performance. This part will improve the torque performance by optimizing the eccentric distance h_e. As shown in Figure 9, fixing O, the eccentric distance h_e varies as O' moves. The minimum air gap is decreasing as h_e is increasing, so the eccentric distance h_e should not be too large in order to ensure the air-gap length.

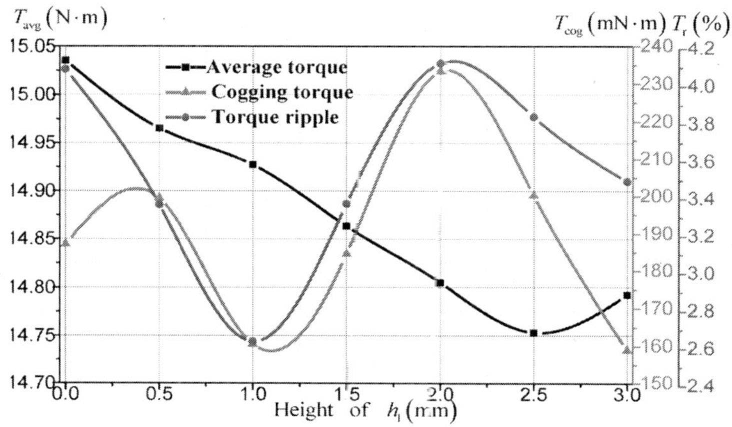

(a)

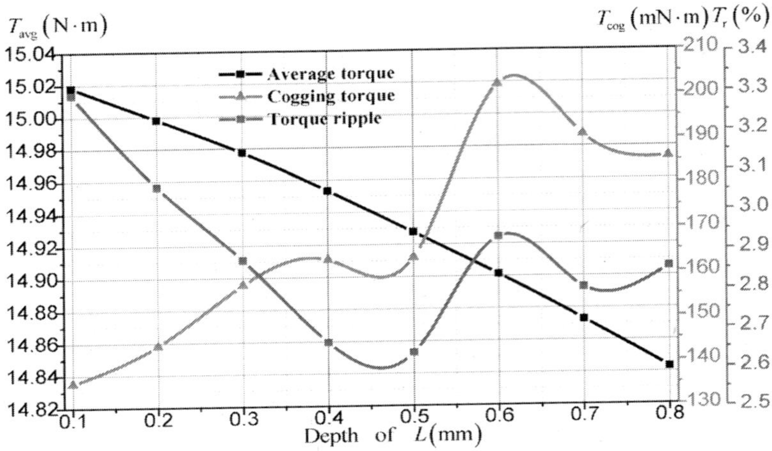

(b)

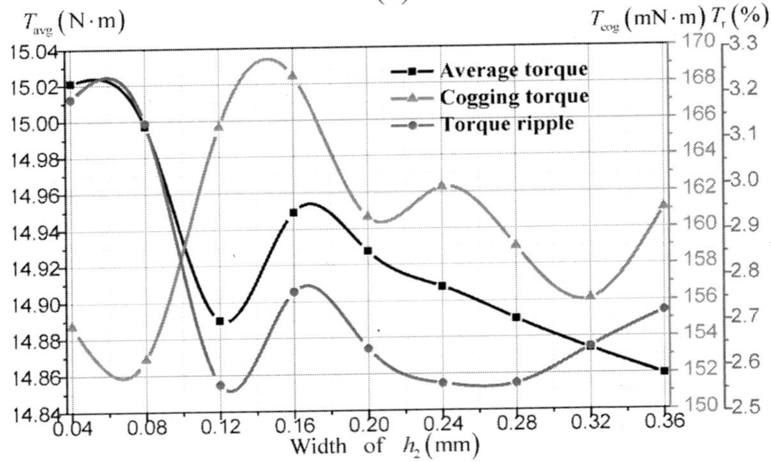

(c)

Figure 6. Average torques, torque ripples, cogging torques according to: (**a**) the height of h_1; (**b**) the depth of L; and (**c**) the width of h_2.

Table 2. Torque characteristics before and after the notched teeth optimization.

Optimizing procedure	Cogging torque (mN·m)	Torque ripple (%)	Average torque (N·m)
Initial structure	187.11	4.09	15.04
Optimized structure with h_1 = 1.0 mm, L = 0.5 mm, and h_2 = 0.28 mm	158.97	2.56	14.89

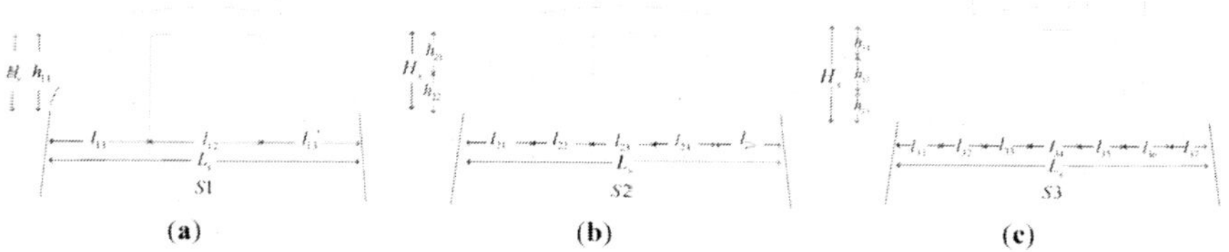

Figure 7. Stepped teeth schemes and design variables: (**a**) S1; (**b**) S2; and (**c**) S3.

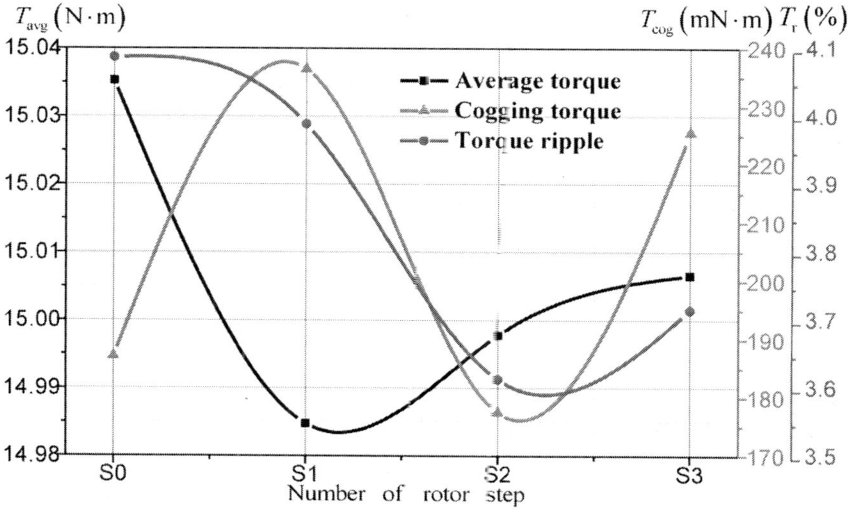

Figure 8. Average torques, torque ripples, cogging torques according to the number of rotor step.

Table 3. Torque characteristics before and after the stepped teeth optimization.

Optimizing procedure	Cogging torque (mN·m)	Torque ripple (%)	Average torque (N·m)
Initial structure	187.11	4.09	15.04
Two step structure	177.45	3.62	14.93

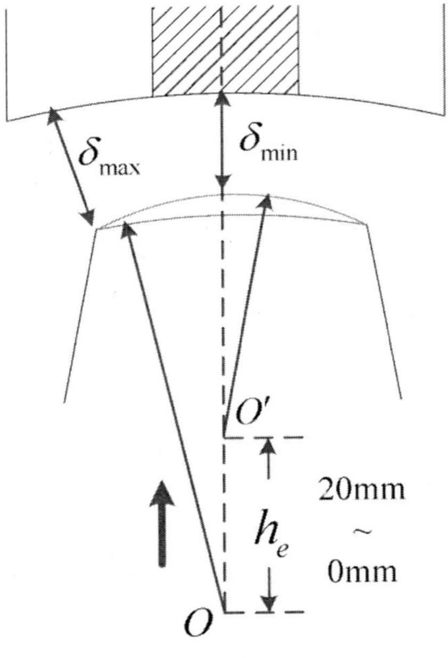

Figure 9. The model of rotor with eccentric teeth.

In this paper, the eccentric distance varies from 0 mm to 20 mm at an interval of 2.5 mm firstly. It is found that the cogging torque and torque ripple are low in the range of 12.5–17.5 mm. Then, a more detailed interval of 0.5 mm is adopted in the range of 12.5–17.5 mm for further research. The results are shown in Figure 10. The torque ripple reaches its minimum value at h_e = 16.5 mm. With the increase of eccentric distance, the electromagnetic torque keeps increasing, while the torque ripple firstly decreases and then increases, and the cogging torque firstly increases, then decreases and finally increases again.

Based on the analysis results, we can draw a conclusion that the eccentric distance h_e of 16.5 mm is more preferred to improve the torque characteristics. Torque characteristics before and after the eccentric teeth optimization are shown inTable 4. It can be seen that the effect of improving the torque characteristics by eccentric teeth is not satisfactory enough.

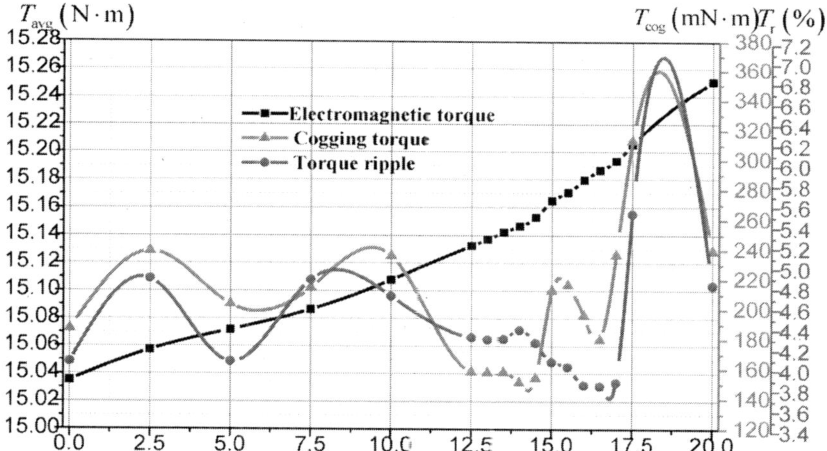

Figure 10. Average torques, torque ripples, cogging torques according to the distance of h_e.

Table 4. Torque characteristics before and after the eccentric teeth optimization.

Optimizing procedure	Cogging torque (mN·m)	Torque ripple (%)	Average torque (N·m)
Initial structure	187.11	4.09	15.04
Optimized structure with h_e = 16.5 mm	180 08	3.85	15.12

3.4. Rotor with the Combination of Stepped and Notched Teeth

The combination of the stepped teeth S2 scheme and the notched teeth scheme is shown in Figure 11. The design process of the v-shape notch is similar to the process shown in Figure 5.

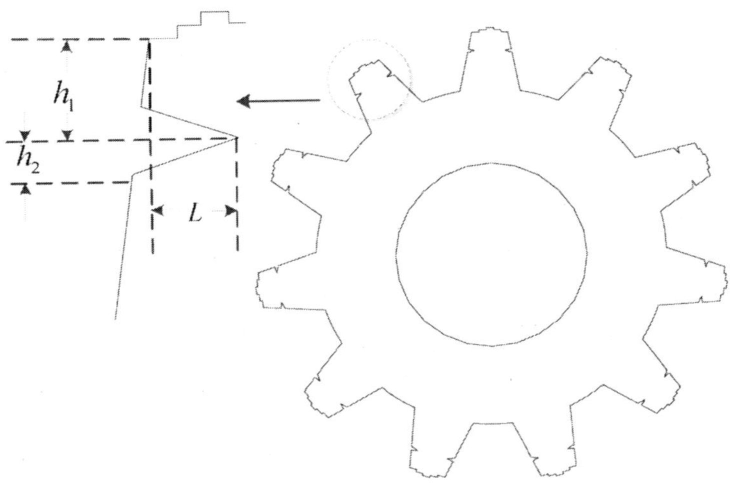

Figure 11. The model of rotor with the combination of stepped and notched teeth.

The changing laws of average electromagnetic torque, cogging torque and torque ripple with the variation of parameters are shown in Figure 12. The torque data of point h_1 = 0 mm in Figure 12a represents the torque data of the stepped teeth S_2scheme.

Based on the compromise principle (the cogging torque and torque ripple are effectively reduced but with an average torque loss as small as possible), the optimal v-shape notch parameters for this composite scheme with h_1 = 1.5 mm, L = 0.4 mm, h_2 = 0.16 mm are selected after carefully comparing the results.

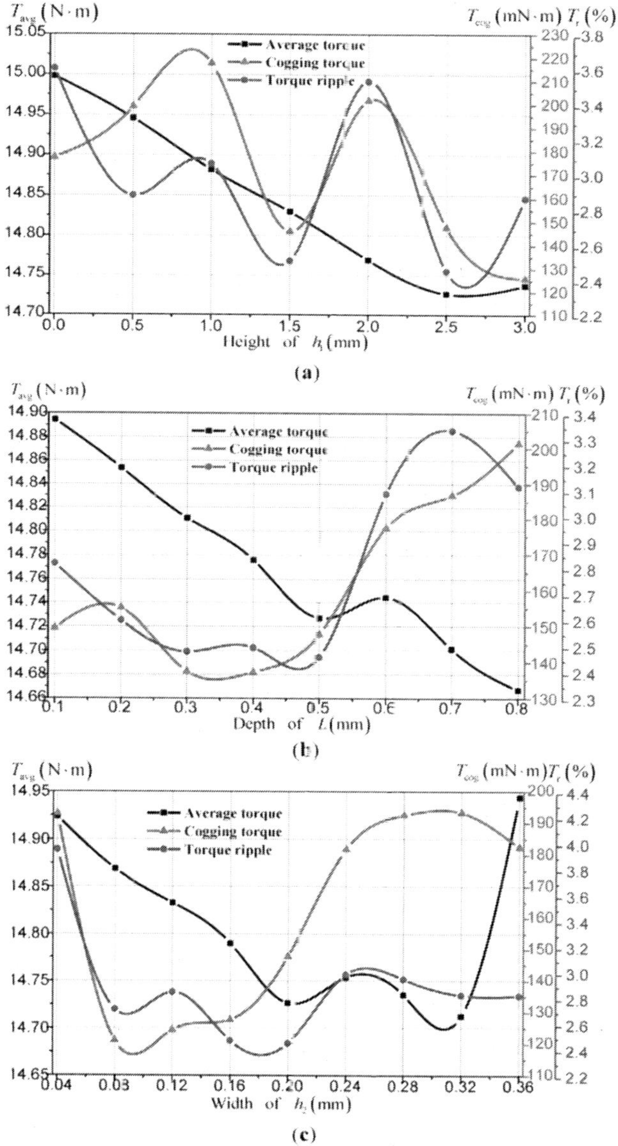

Figure 12. Average torques, torque ripples, cogging torques according to: (**a**) the height of h_1; (**b**) the depth of L; and (**c**) the width of h_2.

Torque characteristics before and after optimizing the stepped and notched teeth combination are compared in Table 5. After adopting the appropriate combination of stepped and notched teeth, the cogging torque and torque ripple are further reduced to 127.65 mN·m and 2.49% by 31.78% and 39.15% with only 1.63% loss of average electromagnetic torque.

Table 5. Torque characteristics before and after optimizing the stepped and notched teeth combination.

Optimizing procedure	Cogging torque (mN·m)	Torque ripple (%)	Average torque (N·m)
Initial structure	187.11	4.09	15.04
Optimized structure with h_1 = 1.5 mm, L = 0.4 mm, h_2 = 0.16 mm, and stepped teeth S2	127.65	2.49	14.79

3.5. Rotor with the Combination of Eccentric and Notched Teeth

The combination of the eccentric teeth and the notched teeth is shown in Figure 13. In this scheme, by fixing O and moving O', the eccentric distance of 16.5 mm is adopted based on the results in Section 3.3. Then, keeping the eccentric distance of 16.5 mm unchanged, the design process of v-shape notches is similar to the process shown in Figure 5.

The changing laws of average electromagnetic torque, cogging torque and torque ripple with the variation of parameters are shown in Figure 14. It should be noted that the data of point h_1 = 0 mm, in Figure 14a represents the torque data with only the eccentric teeth and without notched teeth.

Also based on the compromise principle, the most appropriate parameters of v-shape notches are h_1= 1.0 mm, L = 0.2 mm, h_2 = 0.28 mm, in this combination. Torque characteristics before and after optimizing the eccentric and notched teeth combination are compared in Table 6. Compared to the initial scheme, the cogging torque is just decreased by 10 mN·m, but the torque ripple is greatly reduced by 37.24%, while the average electromagnetic torque is slightly increased by 0.59%.

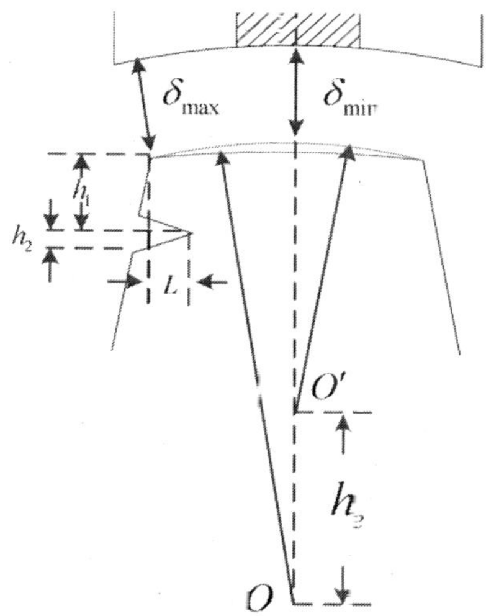

Figure 13. The model of rotor with the combination of eccentric and notched teeth.

4. ROTOR COMPARISON

The rotor with the initial structure, rotor with notched teeth, rotor with stepped teeth, rotor with eccentric teeth, rotor with the combination of stepped and notched teeth, rotor with the combination of eccentric and notched teeth are labeled as T0–T5, respectively. Then the influences of different rotor teeth shapes on the performances are compared in this section.

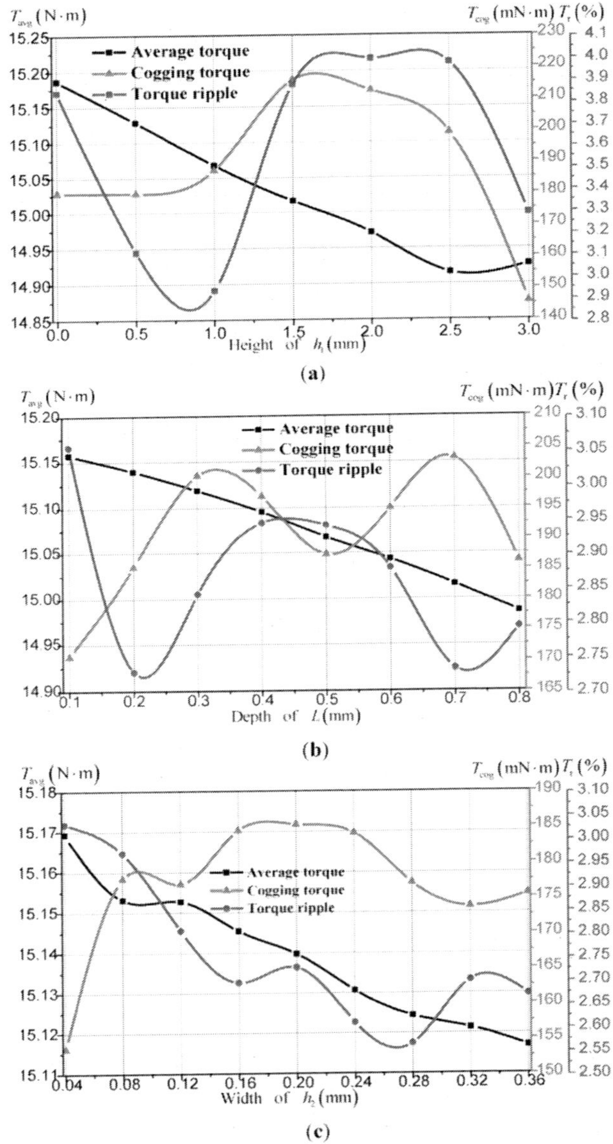

Figure 14. Average torques, torque ripples, cogging torques according to: (**a**) the height of h_1; (**b**) the depth of L; and (**c**) the width of h_2.

Table 6. Torque characteristics before and after optimizing the eccentric and notched teeth combination.

Optimizing procedure	Cogging torque (mN·m)	Torque ripple (%)	Average torque (N·m)
Initial structure	187.11	4.09	15.04
Optimized structure with h_1 = 1.0 mm, L = 0.2 mm, h_2 = 0.28 mm, h_e = 16.5 mm	177.00	2.57	15.12

4.1. Cogging Torque

The cogging torques of different schemes are plotted in Figure 15a, while the maximal and minimal values of the cogging torque are given in Figure 15b. The scheme T4 of the combination of stepped and notched teeth has the lowest cogging torque 127.65 mN·m, followed by the T1 with 158.97 mN·m, T3 with 172.87 mN·m, T2 with 176.13 mN·m, T5 with 177.00 mN·m, while the initial structure T0 with the largest cogging torque 187.11 mN·m.

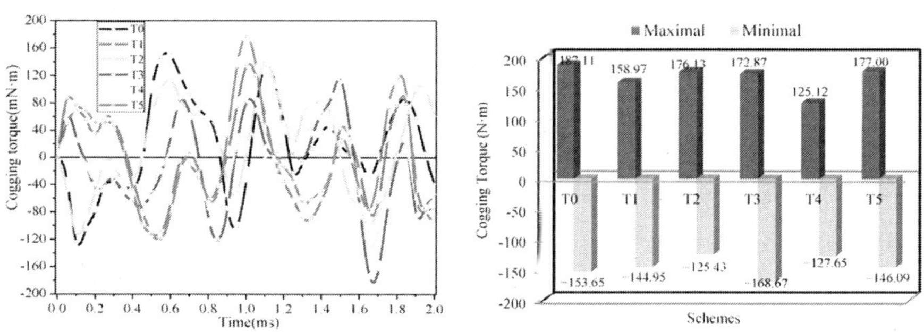

(a) (b)

Figure 15. Cogging torque: (**a**) cogging torque at rated speed, 600 rpm; and (**b**) maximal and minimal values.

4.2. Electromagnetic Torque

The electromagnetic torques of machines with different rotor teeth shapes at 600 rpm excited by 15.4 A current are given in Figure 16a. The statistics on average electromagnetic torques and torque ripples are summarized in Figure 16b. As shown in Figure 16b, the schemes of T3 and T5 have higher electromagnetic torque than the other four schemes due to their smaller air-gap lengths. On the contrary, the schemes of T1, T4 and T5 have better suppression effect on the torque ripple while the electromagnetic torques change little.

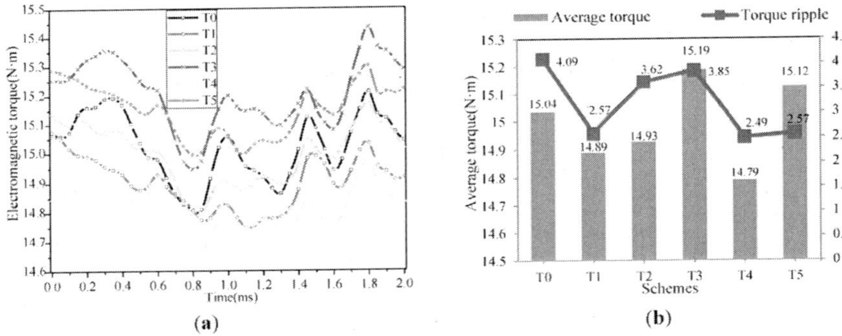

(a) (b)

Figure 16. Torque: (**a**) torque at rated speed, 600 rpm; and (**b**) average values and torque ripple.

4.3. Flux Linkage

The curves of phase-A flux linkage of air-gap excited by the PM are shown in Figure 17. The amplitudes of flux linkage with different rotor teeth shapes are different because the equivalent magnetic resistances of different rotor teeth shapes are varied slightly.

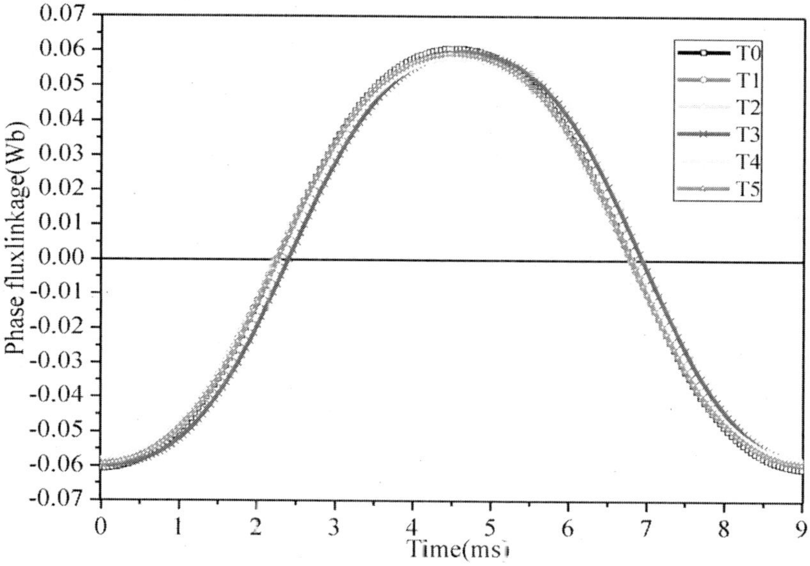

Figure 17. Profiles of flux linkage of Phase A in one electrical cycle.

4.4. Back Electromotive Force

Figure 18a shows the waveforms of no-load back EMFs with different rotor teeth shapes. Because of the non-full sinusoid of the air gap flux density, back EMF waveforms are not sinusoidal enough. The harmonic are analyzed by fast Fourier transform (FFT), as shown in Figure 18b:

$$THD = \frac{100\sqrt{\sum_{k=2}^{K} U_K^2}}{U_1} \tag{4}$$

According to Equation (4) [4], the THD of the initial structure is 14.24%, after adopting the novel schemes, the THD are 15.09% in

T1, 14.02% in T2, 14.90% in T3, 14.25% in T4 and 15.32% in T5, respectively. The rotor with stepped teeth scheme T2 has the best sinusoidal back EMF. As illustrated in Figure 18b, all schemes exhibit higher third harmonic, which can be eliminated by taking advantage of the Y type connection winding of the FSPM machine. Omitting the third harmonic, the harmonic proportions of each scheme are significantly reduced to 1.23% in T0, 1.33% in T1, 1.18% in T2, 1.24% in T3, 1.19% in T4 and 1.24% in T5, respectively. By comparison, the rotor with stepped teeth scheme T2 still has the best sinusoidal back EMF waveform.

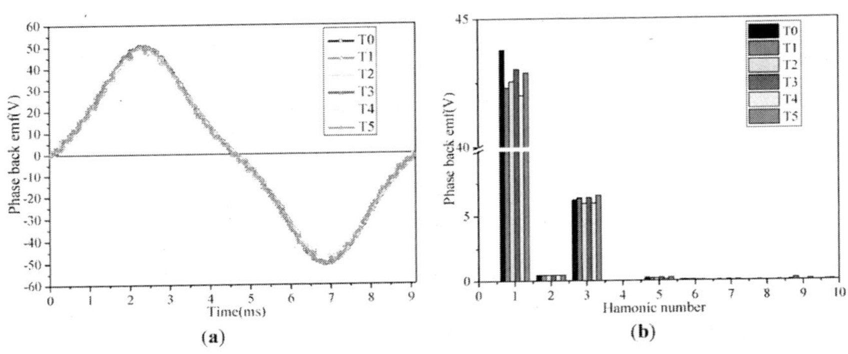

Figure 18. Back electromotive force (EMF): (**a**) profiles in one electrical cycle; and (**b**) harmonics.

4.5. Summary of the Results

The cogging torques, electromagnetic torques, torque ripples, flux linkages, back EMFs and their relative harmonics of the machines with different rotor teeth shapes are summarized in Table 7. Considering the torque characteristics and the waveforms of back EMFs, T4 is the best scheme. It reduces the cogging torque by

31.87% and torque ripple by 39.15% but losses average torque only by 1.63%. Its THD of back EMF is only next to T2.

Table 7. Comparatively comparison of the performances of T0–T5 schemes. THD: total harmonic distortion.

Performance	T0	T1	T2	T3	T4	T5
Cogging torque (mN·m)	187.11	158.97	176.13	172.87	127.65	177.00
Average torque (N·m)	15.04	14.89	14.93	15.19	14.79	15.12
Torque ripple (%)	4.09	2.57	3.62	3.85	2.49	2.57
THD of EMF (%)	14.24	15.09	14.02	14.90	14.25	15.32
THD of EMF without three harmonic (%)	1.23	1.33	1.18	1.24	1.19	1.24

5. CONCLUSIONS

As one of the key part of the EV driveline system, a 12-slot/11-pole FSPM machine used for EVs is researched in this paper. Considering the comparatively larger cogging torque and torque ripple caused by the doubly salient structure, five rotor teeth shapes, *i.e.*, rotor with notched teeth, rotor with stepped teeth, rotor with eccentric teeth, rotor with the combination of stepped and notched teeth, rotor with the combination of eccentric and notched teeth, are investigated to improve the torque characteristics. The laws of torque characteristic changes with the variation of the key parameters in different schemes are obtained and analyzed. The performance with different rotor teeth shapes are compared and discussed comprehensively from the aspects of cogging torque, torque ripple, electromagnetic torque, flux linkage, back EMF, and so on. It is found that the rotor with the combination of stepped and notched teeth has the best effectiveness to reduce the cogging torque by 31.87% and torque ripple by 39.15% but with a slight average torque loss of 1.63%. However, these methods all lead to more or less increased manufacturing costa and some of them may lead to considerable acoustic noise, which will be studied in our future work.

ACKNOWLEDGMENTS

This work was supported in part by the National Natural Science Foundation of China under Project 51307008, in part by the Key laboratory for Intelligent Control & Decision of Complex Systems of Beijing Institute of Technology, in part by Ph.D. Programs Foundation of Ministry of Education of China under Project 20121101120024, in part by Basic Research Foundation of Beijing Institute of Technology under Grant 20110642015, 20120642013 and 20130642015, and in part by Excellent Young Scholars Research Fund of Beijing Institute of Technology.

AUTHOR CONTRIBUTIONS

All authors contributed to this work by collaboration. Jing Zhao was the main author of this manuscript. Yashuang Yan assisted in the design of optimization methods to this study. Bin Li and Zhen Chen provided some useful suggestions in the construction of paper. The whole project was supervised by Xiangdong Liu. All authors revised and approved the publication.

CONFLICTS OF INTEREST

The authors declare no conflict of interest.

REFERENCES

1 Fei, W.Z.; Luk, P.C.K.; Miao, D.-M.; Shen, J.-X. Investigation of torque characteristics in a novel permanent magnet flux switching machine with an outer-rotor configuration. *IEEE Trans. Magn.* **2013**, *45*, 4656–4659.

2 Fei, W.; Luk, P.C.K.; Shen, J.X.; Wang, Y.; Jin, M. A novel permanent-magnet flux switching machine with an outer-rotor configuration for in-wheel light traction applications. *IEEE Trans. Ind. Appl.* **2012**, *48*, 1496–1506.

3 Ahmad, M.Z.; Sulaiman, E.; Haron, Z.A.; Kosaka, T. Preliminary studies on a new outer-rotor permanent magnet flux switching machine with hybrid excitation flux for direct drive EV applications. In Proceedings of the 2012 IEEE International Conference on Power and Energy (PECon), Kota Kinabalu, Malaysia, 2–5 December 2012; pp. 928–933.

4 Hwang, C.-C.; Chang, C.-M.; Hung, S.-S.; Liu, C.-T. Design of high performance flux switching PM machines with concentrated windings. *IEEE Trans. Magn.* **2014**, *50*.

5 Xue, X.; Zhao, W.; Zhu, J.; Liu, G.; Zhu, X.; Cheng, M. Design of five phase modular flux switching permanent magnet machines for high reliability applications. *IEEE Trans. Magn.* **2013**, *49*, 3941–3944.

6 Abdollahi, S.E.; Vaez-Zadeh, S. Back EMF analysis of a novel linear flux switching motor with segmented secondary.*IEEE Trans. Magn.* **2014**, *50*, 1–9.

7 Zulu, A.; Mecrow, B.C.; Armstrong, M. Permanent-magnet flux-switching synchronous motor employing a segmental rotor. *IEEE Trans. Ind. Appl.* **2012**, *48*, 2259–2267.

8 Thomas, A.S.; Zhu, Z.Q.; Owen, R.L.; Jewell, G.W.; Howe, D. Multiphase flux-switching permanent-magnet brushless machine for aerospace application. *IEEE Trans. Ind. Appl.* **2009**, *45*, 1971–1981.

9 Thomas, A.S.; Zhu, Z.Q.; Jewell, G.W. Proximity loss study in high speed flux-switching permanent magnet machine.*IEEE Trans. Magn.* **2009**, *45*, 4748–4751.

10 Sulaiman, E.; Kosaka, T.; Matsui, N. Parameter optimization study and performance analysis of 6S-8P permanent magnet flux switching machine with field excitation for high speed hybrid electric vehicles. In Proceedings of the 2011-14th European Conference on Power Electronics and Applications (EPE 2011), Birmingham, UK, 30 August–1 September 2011; pp. 1–9.

11 Hua, W.; Cheng, M.; Zhang, G. A novel hybrid excitation flux-switching motor for hybrid vehicles. *IEEE Trans. Magn.***2009**, *45*, 4728–4731.

12 Wang, Y.; Sun, J.; Zou, Z.; Wang, Z.; Chau, K.T. Design and analysis of a HTS flux-switching machine for wind energy conversion. *IEEE Trans. Appl. Supercond.* **2013**, *23*.

13 Zhang, G.; Cheng, M.; Hua, W.; Dong, J. Analysis of the oversaturated effect in hybrid excited flux-switching machines. *IEEE Trans. Ind. Magn.* **2011**, *47*, 2827–2830.

14 Raminosoa, T.; Gerada, C.; Galea, M. Design considerations for a fault-tolerant flux-switching permanent-magnet machine. *IEEE Trans. Ind. Electron.* **2011**, *58*, 2818–2825.

15 Li, G.; Ojeda, J.; Hoang, E.; Gabsi, M. Double and single layers fault-tolerant flux-switching permanent-magnet motor: Fault tolerant model for critical applications. In Proceedings of the

2011 International Conference on Electrical Machines and Systems (ICEMS), Beijing, China, 20–23 August 2011.

16 Raminosoa, T.; Gerada, C. Fault tolerant winding technology comparison for flux switching machine. In Proceedings of the 2010 XIX International Conference on Electrical Machines (ICEM), Rome, Italy, 6–8 September 2010; pp. 1–6.

17 Zhao, W.; Cheng, M.; Chau, K.T.; Hua, W.; Jia, H.; Ji, J.; Li, W. Stator-flux-oriented fault-tolerant control of flux-switching permanent-magnet motors. *IEEE Trans. Magn.* **2011**, *47*, 4191–4194.

18 Chen, J.T.; Zhu, Z.Q. Winding configurations and optimal stator and rotor pole combination of flux-switching PM brushless AC machines. *IEEE Trans. Energy Convers.* **2010**, *25*, 293–302.

19 Chen, J.T.; Zhu, Z.Q. Comparison of all- and alternate-poles-wound flux-switching PM machines having different stator and rotor pole numbers. *IEEE Trans. Ind. Appl.* **2010**, *46*, 1406–1415.

20 Zheng, P.; Zhao, J.; Han, J.; Wang, J.; Yao, Z.; Liu, R. Optimization of the magnetic pole shape of a permanent-magnet synchronous motor. *IEEE Trans. Magn.* **2007**, *43*, 2531–2533.

21 Zhao, W.; Lipo, T.A.; Kwon, B. Material-efficient permanent-magnet shape for torque pulsation minimization in SPM motors for automotive applications. *IEEE Trans. Ind. Electron.* **2014**, *61*, 5779–5787.

22 Fei, W.; Luk, P.C.K.; Shen, J. Torque analysis of permanent-magnet flux switching machines with rotor step skewing. *IEEE Trans. Magn.* **2012**, *48*, 2664–2673.

23 Fei, W.; Luk, P.C.K.; Shen, J.X.; Xia, B.; Wang, Y. Permanent-magnet flux-switching integrated starter generator with

different rotor configurations for cogging torque and torque ripple mitigations. *IEEE Trans. Ind. Appl.* **2011**, *47*, 1247–1256.

24 Wang, D.; Wang, X.; Jung, S.-Y. Reduction on cogging torque in flux-switching permanent magnet machine by teeth notching schemes. *IEEE Trans. Magn.* **2012**, *48*, 4228–4231.

25 Abdollahi, S.E.; Vaez-Zadeh, S. Reducing cogging torque in flux switching motors with segmented rotor. *IEEE Trans. Magn.* **2013**, *49*, 5304–5309.

26 Zhu, Z.Q.; Chen, J.T.; Pang, Y.; Howe, D.; Iwasaki, S.; Deodhar, R. Analysis of a novel multi-tooth flux-switching PM brushless AC machine for high torque direct-drive applications. *IEEE Trans. Magn.* **2008**, *44*, 4313–4316.

27 Cai, J.; Lu, Q.; Jin, Y.; Chen, C.; Ye, Y. Performance investigation of multi-tooth flux-switching PM linear motor. In Proceedings of the 2011 International Conference on Electrical Machines and Systems (ICEMS), Beijing, China, 20–23 August 2011.

28 Xu, W.; Lei, G.; Wang, T.; Yu, X.; Zhu, J.; Guo, Y. Theoretical research on new laminated structure flux switching permanent magnet machine for novel topologic plug-in hybrid electrical vehicle. *IEEE Trans. Magn.* **2012**, *48*, 4050–4053.

29 Zhao, W.; Cheng, M.; Chau, K.T.; Cao, R.; Ji, J. Remedial injected-harmonic-current operation of redundant flux-switching permanent-magnet motor drives. *IEEE Trans. Ind. Electron.* **2013**, *60*, 151–159.

30 Zhao, W.; Cheng, M.; Hua, W.; Jia, H.; Cao, R. Back-EMF harmonic analysis and fault-tolerant control of flux-switching permanent-magnet machine with redundancy. *IEEE Trans. Ind. Electron.* **2011**, *58*, 1926–1935.

31 Zhao, W.; Cheng, M.; Cao, R.; Ji, J. Experimental comparison of remedial single-channel operations for redundant flux-switching permanent magnet motor drive. *Prog. Electromagn. Res.* **2012**, *123*, 189–204.

CHAPTER 10

Development of a Micro-Step Voltage-Fed Actuator with a Novel Stepper Motor for Automobile AGS Systems

Se-Hyun Rhyu [1,*], Jeong-Jong Lee [1], Bon-Gwan Gu [2], Byung-Dae Choi [3] and Jung-Hyuk Lim [3]

[1] Korea Electronics Technology Institute, 203-101, Bucheon TP 192 Yakdae-dong, Wonmi-gu, Bucheon-si, Gyeonggi-do 420-140, Korea

[2] School of Energy Engineering, Kyungpook National University, Daegu 702-701, Korea

[3] O&K Technology Co. Ltd., Seoul 152-769, Korea

ABSTRACT

This paper presents an improved micro-step voltage-fed actuator for an automobile active grill shutter (AGS) system. A novel structured stepper motor, which contains both the main and auxiliary teeth in the stator, is proposed for the actuator. In a normal permanent magnet (PM) motor coils are generally wound on all the stator teeth, however, in the proposed motor, the winding is only on the main teeth. Because of the absence of coils in the auxiliary teeth, the proposed stepper motor possesses the following advantages: simple structure, lighter weight, smaller volume, and less time consumption. The unique auxiliary poles in the stepper motor supply the flux path to increase the step resolution even without any coils. The characteristics of the proposed stepper motor were investigated using finite element analysis. In particular, the effect of the magnetization distribution of the PM on the motor performance was investigated during the analysis. Cogging torque, which causes noise and vibration issues, was minimized by the tooth-shape optimization. In addition, a micro-step voltage-fed algorithm was implemented for a high-resolution position control. By employing a current close to a sine wave using space vector pulse-width modulation, a high-quality current waveform with a high resolution was obtained. Finally, the proposed prototype was fabricated, and the cogging torque, back-electromotive force, and current characteristics were measured by mounting the prototype on the AGS system. Both the analysis and experimental results validate the performance improvement from the proposed motor and its possible application for the flap control of the AGS system.

KEYWORDS

stepper motor; active grill shutter (AGS); micro-step; actuator; finite element analysis (FEA); cogging torque; space vector PWM

1. INTRODUCTION

Recently, active grill shutters (AGSs) have been put in the spotlight as actuating modules for fuel economy in automobiles. The actuating module controls the air flap of an AGS system because it can reduce air resistance and improve the cooling effect, which conserves energy during driving by allowing the automobile to run at high speed and thus longer distance. Therefore, the electrical performance of the actuator, which controls the flap angle of the AGS system, is very important. In particular, a light and slim actuator design is required to enhance the fuel economy, and a network-driver-integrated structure design is needed to communicate with the car engine control unit. A gear module with a high deceleration ratio using a slim motor is also required to provide more strength. Thus, the motor should have a very simple structure for it to be installed into the AGS system driving actuator and for proper arrangement of the other parts. Harmful factors, e.g., cogging torque, which cause vibration or decrease the angle control precision of the actuator when driven, should be reduced to the minimum.

Generally, brush-type DC motors are used in AGS systems; however, they reduce the AGS performance because of their low efficiency and weak controllability. On the other hand, the stepper motor is suitable as a control device, such as in the AGS system that requires maximum angular accuracy, because it is capable of open loop control [1]. The step-motor controller can smoothly perform positioning control. To realize smooth positioning, the micro-step mode that uses a pulse-width modulation (PWM) approach is an excellent selection [2,3]. However, designing a small-size slim stepper motor with high performance is difficult because of the complex winding structure and the difficulty of multi-polarization in small-size PMs. In the present paper, a newly structured PM-type stepper motor with auxiliary poles and partial winding is proposed for an automobile AGS system. In addition, we analyze the magnetic characteristics and compare the result with the general-type model using finite element analysis (FEA) to account for the magnetization distribution of the PM in the magnetizing yoke. The simulation and experimental results show that the proposed model can be applied in the AGS system.

2. AGS ACTUATOR AND CHARACTERISTIC ANALYSIS

2.1. Actuator with Novel Stepper Motor for AGS System

Figure 1 shows the air-flow concept according to the flap *ON* and *OFF* in an automobile AGS system. The AGS system has an air flap and an actuating module that consists of a precision motor, a driver, and a gear stack. If an automobile is driven at high speed, the flap is closed by the actuating module. Figure 2 shows the structure of the general type and the proposed three-phase PM-type stepper motor models. The stator consists of three-phase exciting coils and a core with a typical tooth shape. However, relative to the structure among the PM, coil, and tooth width, the coil turns are restricted to reduce the overall PM size following the decrease in the slot width. Furthermore, we confirm that the winding connection and the manufacturing process of the motor are complex, as shown in Figure 2a. In this paper, we propose a slim-type stepper motor with a novel structure with auxiliary poles. Although the proposed model consists of the same parts as the general-type model, it has a different structure. The proposed model is composed of three poles with coils, which can provide the same step resolution because it has three auxiliary poles that offer a magnetic coupling path, as shown in Figure 2b. In addition, the proposed model has an extremely simple winding connection and manufacturing process compared with the general-type model.

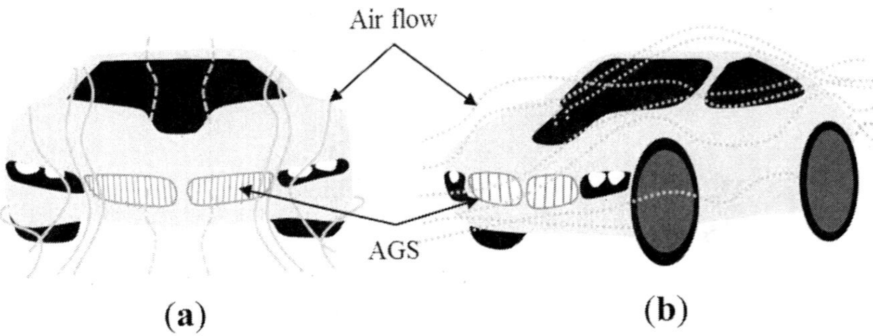

Figure 1. Air-flow concept in AGS. (**a**) Flap *ON*. (**b**) Flap *OFF*.

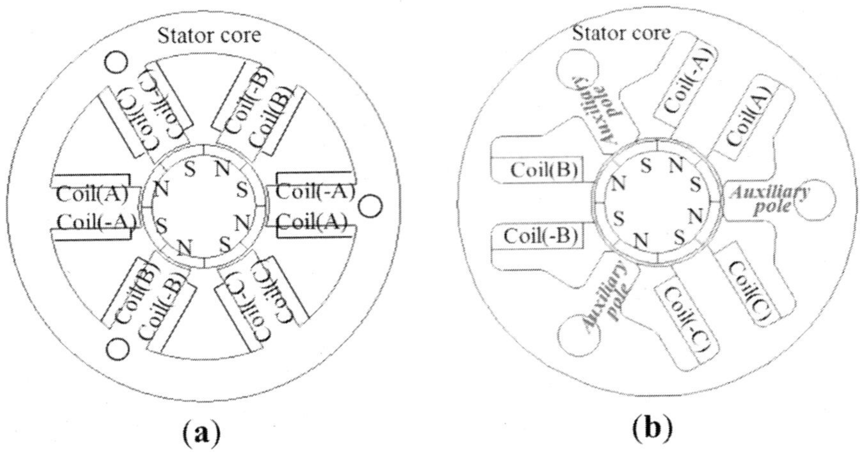

Figure 2. Stator-core shape of the AGS. (**a**) General motor type. (**b**) Proposed motor.

The magnetic characteristics of the proposed motor are analyzed by FEA to account for the PM magnetization. Table 1lists the specifications of the proposed model.

Table 1. Specifications of the proposed stepper motor.

Item	Value	Unit
Number of phase	3	phases
Outer diameter of stator	38.0	mm
Length of stator	4.0	mm
PM type	Bonded NdFeB	-
Number of poles	8	poles

2.2. Characteristic Calculation Using FEA

Figure 3 shows the flowchart of the static and dynamic characteristic analysis using the FE method directly. First, the distribution of the PM magnetic flux density is computed by analyzing the magnetizing yoke. Using the distribution of the magnetization in the PM region, we calculate the magnetic characteristics of the stepper motor [4]. The fundamental equation for the PM magnetic field using the magnetic vector potential **A** is:

$$rot(\upsilon rot\mathbf{A}) = \mathbf{J}_o + \upsilon_o rot\mathbf{M} \tag{1}$$

where υ is the reluctivity, υ_o is the reluctivity of vacuum, $\mathbf{J}_o$ is the exciting current density, and **M** is magnetization of the PM.

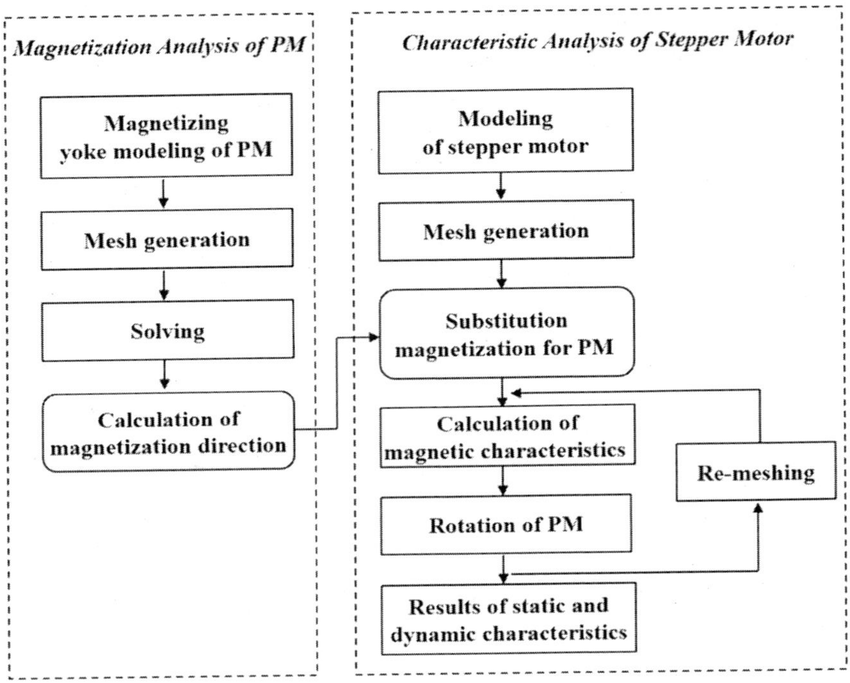

Figure 3. Flowchart of the magnetic characteristic calculation of the stepper motor.

Figure 4 shows the flux line distribution according to the rotor position, which shows that the auxiliary poles also play a key role in the entire magnetic circuit. Figure 5a shows the initial model of a PM-type stepper motor with a normal stator structure. The proposed model is developed on the basis of this model, which has a magnetic circuit in the auxiliary poles, as shown in Figure 5b. In this study, the tooth width and shape are selected as design variables to minimize the cogging torque, which is an important factor in the noise and vibration of the stepper motor. The tooth width and shape acutely affect the main flux and are major factors that determine the size of the coil winding assembly. Figure 6a shows the analysis results of the cogging torque within the range of the tooth width with a 0.5-mm interval in the initial model. Figure 6bshows the cogging torque of the proposed model with a 0.1-mm interval on the basis of the model with a 4-mm tooth width. Here, the amplitude and period of the cogging torque are varied by varying the tooth width, where a minimum value is obtained at a 4.1-mm tooth width. Thus, the model with a 4.1-mm tooth width is considered as the optimized model, which has a cogging torque of 0.45 mN·m, as shown in Figure 7a. Figure 7a shows the evaluation results of the cogging torque, which are the minimized and sinusoidal waveforms obtained from the tooth width and shape variation. For cogging torque reduction and coil inserting, we have cut the tooth shoe using trial and error.

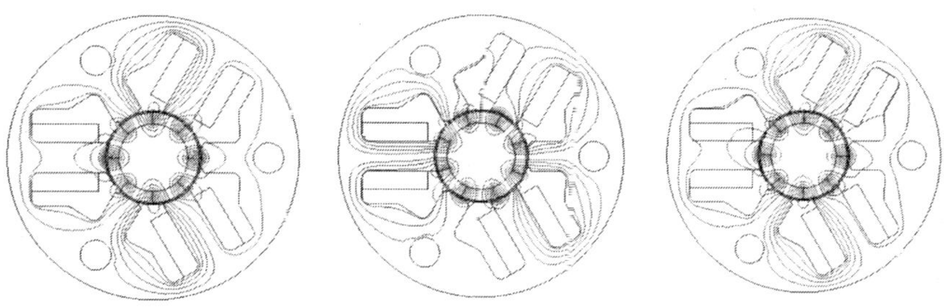

Figure 4. Flux line according to the rotor position: (**a**) 0°, (**b**) 22.5°, and (**c**) 45°.

We must note that the characteristic of the cogging torque is a very important factor for micro-step resolution. In addition,Figure 7b shows the simulation result of the back electromotive force (back-EMF) waveform at 720 rpm, which is a sinusoidal waveform. Thus, we can predict that the cogging torque, which is an important factor to determine, can be possibly reduced by further experiments by considering the exact magnetization distribution of the PM. If the magnetization distribution of the PM is not accurately considered, then many errors would appear on the static and dynamic simulation results. The cogging-torque characteristic factor determines whether the cogging torque can have a minimized amplitude and an optimized sinusoidal waveform by tooth-width and shape optimization. Figure 7b shows that the simulation result waveform of the back-EMF is almost sinusoidal at 720 rpm. Thus, the cogging torque can be possibly reduced. The cogging torque is an important parameter in determining whether the PM is accurately manufactured and its magnetization accurately distributed. Inaccurate PM magnetization distribution would lead to many errors in the static and dynamic simulation results.

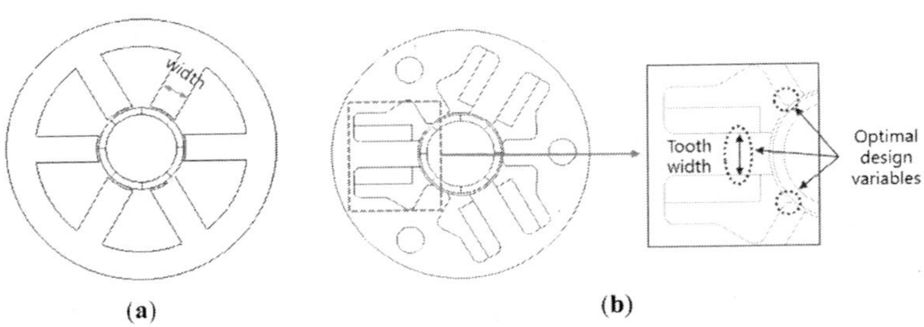

Figure 5. Design variables for the cogging-torque reduction: (**a**) initial model and (**b**) proposed model.

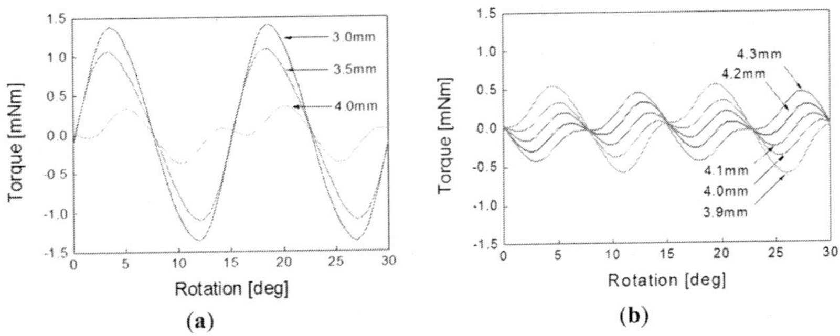

Figure 6. Simulation results of the cogging torque according to the tooth shape: (**a**) initial model and (**b**) proposed model.

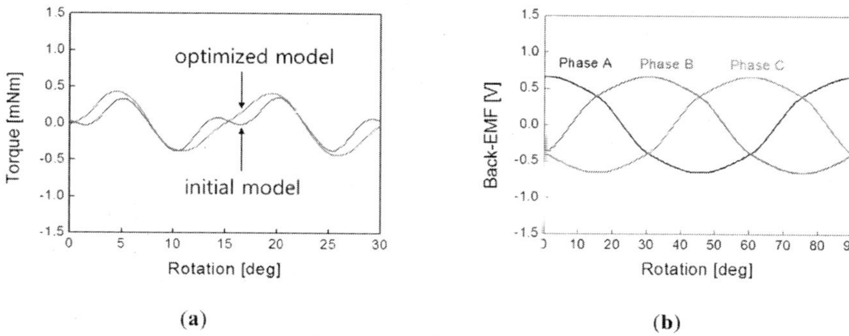

Figure 7. Simulation results of the optimized model: (**a**) cogging torque and (**b**) back-EMF (720 rpm).

3. EXPERIMENTATION

Figure 8 shows the prototypes of the actuating module composed of a stepper motor, a motor driver, and a gear module. The figure shows that the stepper motor, a key driver of the actuator, simply consists of the stator, which consists of the coils, and the rotor, which is a permanent magnet. In particular, while the stator core has six teeth, it has a three-phase winding structure with only three coil groups. The coil of each phase has a structure assembly that allows it to be conveniently plugged at the inner side of the stator. As a result, the torque, speed, and rotational position characteristics of the rotor are output through the gear module to the final shaft of the actuator.

The experimental result of the back-EMF is shown in Figure 9. The amplitude of the back-EMF is approximately 0.69 V. The figure shows that the simulation and experimental results have very similar values. From these results, we can confirm that the PM magnetization is accurately considered in the FEA.

Figure 8. Actuator prototypes of the AGS system: (**a**) upper side and (**b**) lower side.

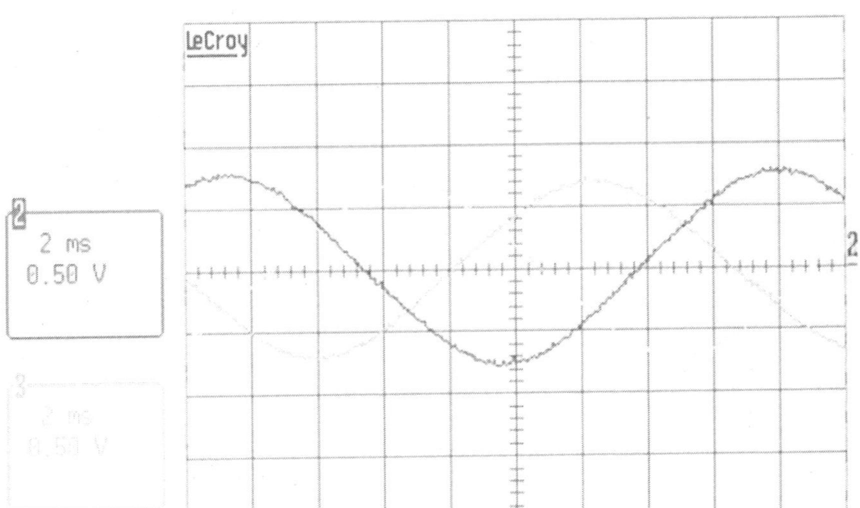

Figure 9. Experimental results of the back-EMF measurement (720 rpm).

Figure 10 shows the experimental setup to measure the characteristics of the cogging torque of the stepper motor using a torque meter (ATM-5MN, SUGAWARA, Tokyo, Japan). After the stepping motor is connected to the torque meter, the torque is measured according to the rotor angle. The measurement of the cogging-torque waveform is shown in Figure 11. Compared with the calculated value of the holding torque shown in Figure 7a, the amplitude of the measured torque is 0.44 mN·m. Both results are very similar and have only a slight error. Also, Table 2 shows the characteristic comparison between developed prototypes and traditional actuator. We can see that the developed actuator has a small size and very light weight compare with commercial model.

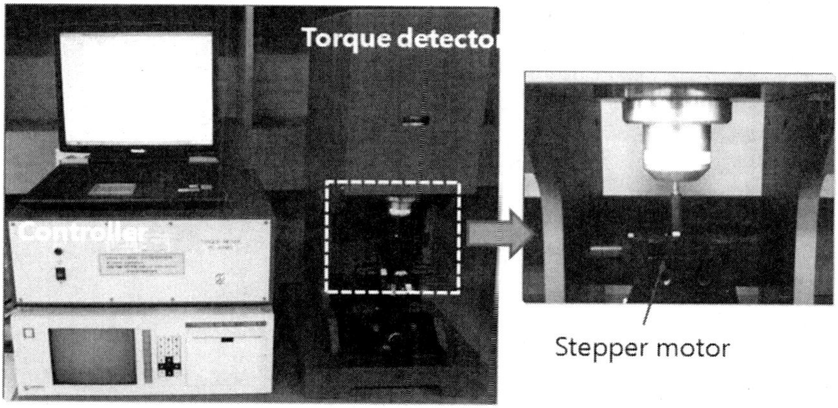

Figure 10. Experimental setup for the cogging-torque measurement.

The PWM method, which controls the magnitude of the voltage, is generally used to control the stepper motor. On the other hand, sinusoidal pulse width modulation (SPWM) is widely used, which simply compares the output voltage and the sine table. The present study used the space vector PWM, which has the advantage of higher differential output voltage, and not the SPWM method to control the voltage. The space vector PWM has an approximately 15% higher voltage magnitude than the SPWM. It can make the voltage differences among the motor lines appear as a sine wave only and has the advantage of being able to reduce the switching loss because it always prevents the occurrence of one-phase switch change [5].

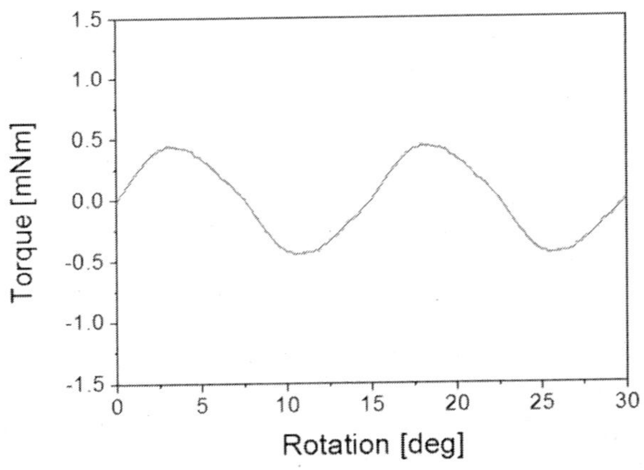

Figure 11. Experimental results of the cogging-torque measurement.

Figure 12 shows the comparison of the PWM duty generation methods. ν∗a, ν∗b and ν∗c are the output voltage commands, S_a, S_b and S_c are switch *ON* and *OFF* signal, respectively. V_{dc} denotes the controller input DC voltage. Figure 12a shows the signal block diagram of the general PWM utilizing the triangle intersection technique. Figure 12b shows the general SPWM method based voltage commands, which generates a PWM signal by simply comparing the output voltage from the sine table with the triangular wave, and the three-phase switch continuously alternates between *ON* and *OFF*. In contrast, the differences among the voltages are maintained, and the phase with the lowest output voltage is reduced to zero, which stops the one-phase switching; thus, the margin of the output voltages is obtained by the space vector PWM, as shown inFigure 12c.

Table 2. Comparison of actuator (commercial model *vs.* developed model).

Item	Commercial Product	Developed Prototype
Type of motor	Brush type DC	Stepper
Size of motor	(*Diameter*) 27.7 mm × (*Length*) 58 mm	(*Diameter*) 38.0 mm × (*Height*) 4.0 mm
Size of actuator	(*Length*) 94 mm × (*Width*) 77 mm × (*Height*) 31 mm	(*Length*) 99.3 mm × (*Width*) 56.2 mm × (*Height*) 23.7 mm
Control method	Open and Shot (*ON* or *OFF*)	Position control
Weight of actuator	285 g	75 g
Photo		

Figure 13 shows the measured voltage and current waveforms of the fabricated actuator prototypes with a micro-step voltage fed control at 3000 rpm and a 1.3-mN·m load. Figure 13 shows that a sinusoidal current waveform with very little harmonic components is generated. In other words, we can conclude that it is very advantageous in terms of vibration or noise when the motor is operated. One cycle of the current waveform has a 180-step resolution, and we confirm the possibility for precision position control owing to the high resolution.

We determine that this favorable result is due to the cogging-torque-reduction design and the sinusoidal back-EMF waveforms of the proposed stepper motor. The precision position control of the flap in the AGS system is very important to control the optimum amount of air flow into the car, and we confirm that the stepper motor and micro-step voltage-fed control method proposed in this paper can be very useful in the flap control of the AGS system.

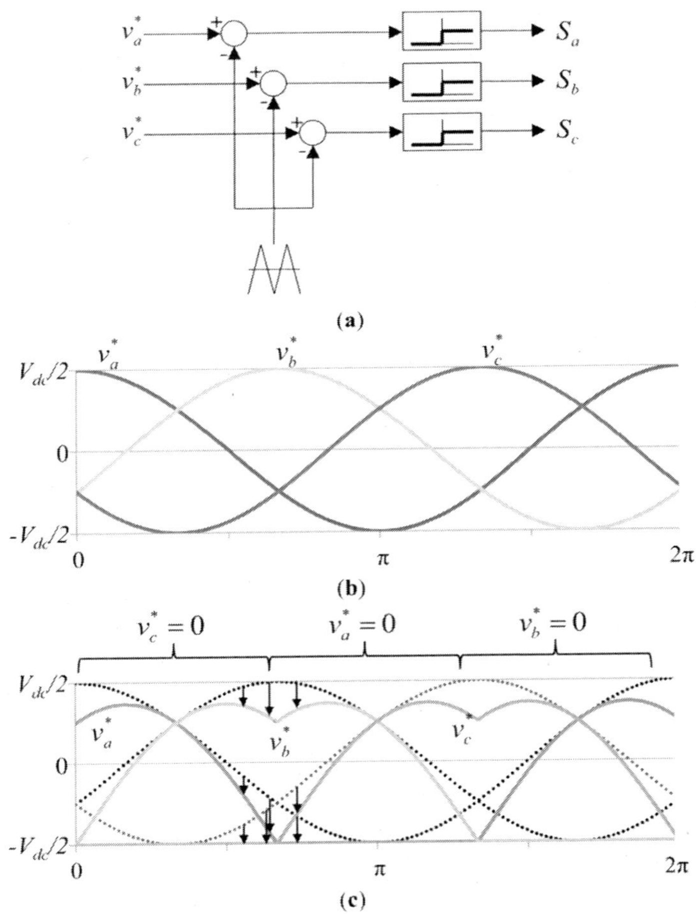

Figure 12. Comparison of the input voltage control methods for the stepper motor. (**a**) Signal block diagram of the PWM. (**b**) Modulation waveforms of SPWM. (**c**) Modulation waveforms of Space vector PWM.

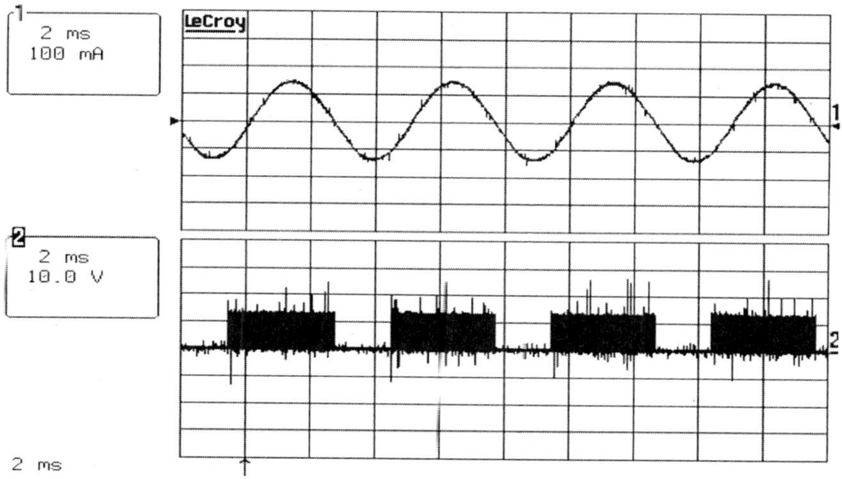

Figure 13. Experimental results of the current and voltage waveforms (3000 rpm, 1.3 mNm).

4. CONCLUSIONS

In this paper, a slim-type stepper motor with a novel structure has been proposed for the automobile AGS system. The shape design of the initial model was changed to a common structure that considers the ease and convenience of production. In particular, a structure that can reduce the number of coils using auxiliary poles has been proposed. The validity of the structural design has been confirmed by the adequacy of the back-EMF waveform and the path of the flux line at each rotor position. Here, we confirmed that the auxiliary pole plays an adequate role as a common pole in the construction of a magnetic circuit. The cogging torque was also minimized by the tooth-shape optimization design, which improved the position-control characteristic. The FEA result that considers the magnetization distribution of the PM was used for the characteristic analysis method. We confirmed through experiments that the cogging torque and the back-EMF characteristics agree well with the analyzed values, and the actuator fabricated with the proposed motor exhibits a high-resolution behavior by measuring the voltage and current waveforms under the micro-step control condition using the space vector PWM.

On the basis of the smaller actuators and high-resolution position, we have developed a novel actuator structure that is easy to manufacture. Thus, we confirmed that our proposed novel stepper motor with partial winding is very useful for actuators used to drive a car AGS system. However, further study related to the position control responsiveness according to the load variation is necessary.

AUTHOR CONTRIBUTIONS

Se-Hyun Rhyu contributed to the computational modeling of the novel stepper motor and coordinated with the manuscript writing. Jeong-Jong Lee contributed to the FE analysis of the stepper motor. Bon-Gwan Gu focused on the formulation of the micro-step-fed control algorithms. Byung-Dae Choi and Jung-Hyuk Lim contributed to the fabrication of the actuator prototypes. All authors contributed to the design and experiments of the actuator, drafting, and revision of the manuscript. All authors approved the final version of this article.

ACKNOWLEDGMENTS

This research was supported by the Small Business Innovation Development Project Program through the Small and Medium Business Administration (SMBA) in 2013 (No. S2092079).

CONFLICTS OF INTEREST

The authors declare no conflict of interest.

REFERENCES

1 Acarnley, P. *Stepping Motors: A Guide to Modern Theory and Practice*, 4th ed.; IEE Control Engineering Series 63; IET: Michael Faraday House, London, UK, 2002.

2 Baluta, G. Microstepping mode for stepper motor control. Proceedings of 2007 International Symposium on Signals, Circuits and Systems (ISSCS), Iasi, Romania, 13–14 July 2007; Volume 2, pp. 1–4.

3 Lee, D.-H.; Che, W.; Ahn, J.-W. Micro-step position control with a simple voltage controller using low-cost micro-processor. Proceedings of 2010 IEEE International Symposium on Industrial Electronics (ISIE), Bari, Italy, 4–7 July 2010; pp. 1378–1382.

4 Rhyu, S.-H.; Kwon, B.-I. Newly structured micro permanent magnet type stepping motor with claw-poles. *IEEE Trans. Magn.* **2006**, *42* (Suppl. 4), 1331–1334.

5 Hava, A.M.; Sul, S.K.; Kerkman, R.J.; Lipo, T.A. Dynamic overmodulation characteristics of triangle intersection PWM methods. *IEEE Trans. Ind. Appl.* **1999**, *35* (Suppl. 4), 896–907.

CHAPTER 11

A Novel Angular Acceleration Sensor Based on the Electromagnetic Induction Principle and Investigation of Its Calibration Tests

Hao Zhao [1] **and Hao Feng** [2,*]

[1] Department of Mechanical and Electrical Engineering, Jiaxing University, No.56, Yuexiu South Road, Jiaxing 314001, China

[2] Department of Information and Engineering Life, Hangzhou Dianzi University, No.1 Avenue, 2 Xiasha, Jianggan District, Hangzhou 310018, China

ABSTRACT

An angular acceleration sensor can be used for the dynamic analysis of human and joint motions. In this paper, an angular acceleration sensor with novel structure based on the principle of electromagnetic induction is designed. The method involves the construction of a constant magnetic field by the excitation windings of sensor, and the cup-shaped rotor that cut the magnetic field. The output windings of the sensor generate an electromotive force, which is directly proportional to the angular acceleration through the electromagnetic coupling when the rotor has rotational angular acceleration. The mechanical structure and the magnetic working circuit of the sensor are described. The output properties and the mathematical model including the transfer function and state-space model of the sensor are established. The asymptotical stability of the sensor when it is working is verified by the Lyapunov Theorem. An angular acceleration calibration device based on the torsional pendulum principle is designed. The method involves the coaxial connection of the angular acceleration sensor, torsion pendulum and a high-precision angle sensor, and then an initial external force is applied to the torsion pendulum to produce a periodic damping angle oscillation. The angular acceleration sensor and the angle sensor will generate two corresponding electrical signals. The sensitivity coefficient of the angular acceleration sensor can be obtained after processing these two-channel signals. The experiment results show that the sensitivity coefficient of the sensor is about 17.29 mv/Krad·s^2. Finally, the errors existing in the practical applications of the sensor are discussed and the corresponding improvement measures are proposed to provide effective technical support for the practical promotion of the novel sensor.

KEYWORDS

sensor; angular acceleration; electromagnetic induction; mathematical model; calibration; error analysis

1. INTRODUCTION

Angular acceleration is a common physical quantity in a rotating system, which can reflect the vibrational state of the rotation angle for the rotation shaft and transmission equipment while working. The responses of the rotation shaft and transmission equipment to various stimulations can be analyzed by measuring the angular acceleration, and especially the dynamic interference of rotary systems can be displayed in the form of angular acceleration. The measurement of rotating mechanical angular acceleration has wide applications, including the fields of automobiles, the military, aerospace, industry, electronics, *etc.* [1–5].

Angular acceleration can be measured indirectly by employing rotation angle sensors or angular speed sensors. However, the signal processing of these methods are very troublesome. Especially, the problems of delay characteristics and noise amplification are hard to resolve, which has prompted the development of direct angular acceleration measurement techniques. The applications of linear acceleration sensor are very wide [6–9], although it is more urgent to develop an angular acceleration sensor without the range limitation of the rotation angle. There exist the following problems in the presented research and development of angular acceleration sensors: lack of range limitation of the rotation angle, small phase lag, high signal-to-noise ratio and a wider band [10].

In recent years, some experts and scholars have conducted a series of researches on angular acceleration measurement. Moody [11] proposed a new sensitive angular accelerometer, which contained a superconducting test mass suspended by a weak flexure pivot; a superconducting quantum interference device amplifier and superconducting circuit were used to detect the angular displacement of the test mass. Sparks [12] described an electroformed micro-machining technology capable of producing both angular rate sensors and accelerometers, which were fabricated by CMOS and provided improved signal output. Jianli Li [13] proposed a novel micro electromechanical system (MEMS) pendulum angular accelerometer with electrostatic actuator feedback; it adopted a proof pendulum with optimized moment of inertia, suspended to dual anchors by a pair of torsion spring beams, as sensing component. Tomikawa [14] dealt with a piezoelectric angular acceleration sensor which can accurately distinguish a signal of rotational motion from other linear motion ones. In order to overcome the influence of gravity in angular acceleration measurements, a system for knee joint motions was constructed with a couple of accelerometers by Yamamoto [15]. The accurate angular acceleration could be calculated as the difference between two output values of the accelerometers. A bulk silicon micro-machined structure 6.4 mm × 6.4 mm in size designed and fabricated by Mizuno could detect angular acceleration and acceleration simultaneously and independently [16]. An angular acceleration measurement system without moving parts using a PVDF and a piezo-composite presented by Marat-Mendes had very low cost compared with other sensors [17]. A novel micro-fluidic channel angular accelerometer was proposed by Wolfaardt, and the sensor with a resolution of 15 mu rad/s^2 and a bandwidth of 50 Hz was demonstrated [18].

In this paper, an angular acceleration sensor without the range limitation of the rotation angle is proposed to measure the instantaneous angular acceleration of the rotating system directly, based on the principle of electromagnetic induction. In addition, the installation of the sensor is very convenient, as the sensor just needs to be coaxially connected with the rotating system when it is used. The output signal of the sensor can be led directly without the slip ring to avoid the weakening of the signal. This paper describes the mechanical structure and working principle of this angular acceleration sensor, and constructs the mathematical model of the sensor, including the transfer function and state-space model. The asymptotical stability of the sensor is verified by the Lyapunov theorem when the sensor is working. Finally, the sensor calibration is demonstrated, and the probable existing errors and their impact on the applications of the sensor are analyzed.

2. THE STRUCTURE AND PRINCIPLE OF THE SENSOR

2.1. The Mechanical Structure of Sensor

The 3D assembly diagram of the angular acceleration sensor based on electromagnetic induction is shown in Figure 1, which includes front cover ①, bearing ②, sensor shaft ③, cup-shaped rotor ④, case ⑤, outer stator core ⑥, output windings ⑦, excitation stator core ⑧, excitation windings ⑨, as well as the end cover ⑩. The excitation stator core ⑧ is composed of a high magnetic permeability soft magnetic iron-nickel alloy sheet or high magnetic permeability silicon steel through punching and shearing, and the excitation windings ⑨ are embedded in its grooves; excitation windings ⑨ are used for the generation of the constant air-gap magnetic field; the cup-shaped rotor ④ is a thin-walled non-magnetic cup, which is prepared from high resistivity phosphor bronze, silicon manganese bronze and tin zinc bronze and other materials; the sensor shaft ③ is connected coaxially with the shaft of the measured rotating system when measuring angular acceleration, and cutting the air-gap magnetic field formed by the excitation windings ⑨; the outer stator core ⑥ is made of the same materials as the excitation stator core ⑧, and output windings ⑦ are embedded in its grooves; the output windings ⑦ will generate the electrical signal which corresponds to the angular acceleration of the rotating testing system. The corresponding 3D sectional view of the sensor is shown in Figure 2, and the sectional sketch map of the sensor is shown inFigure 3.

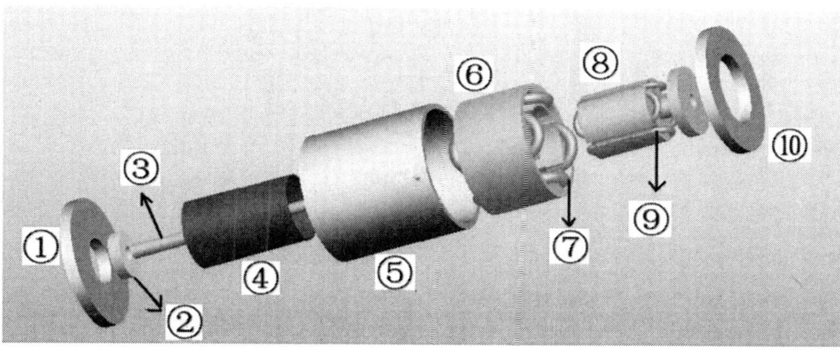

Figure 1. 3D assembly diagram of the angular acceleration sensor.

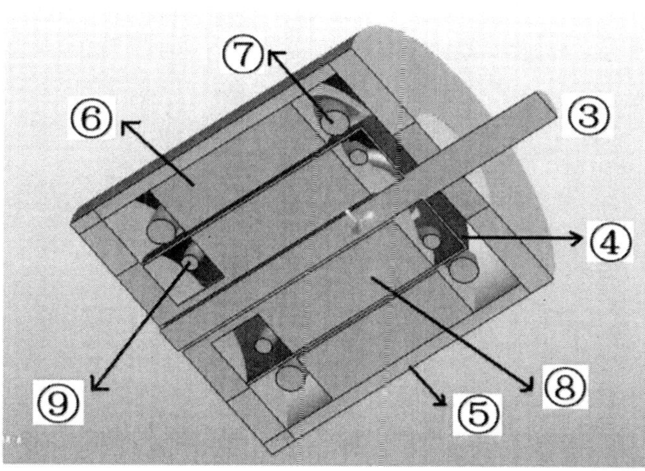

Figure 2. 3D sectional view of the sensor.

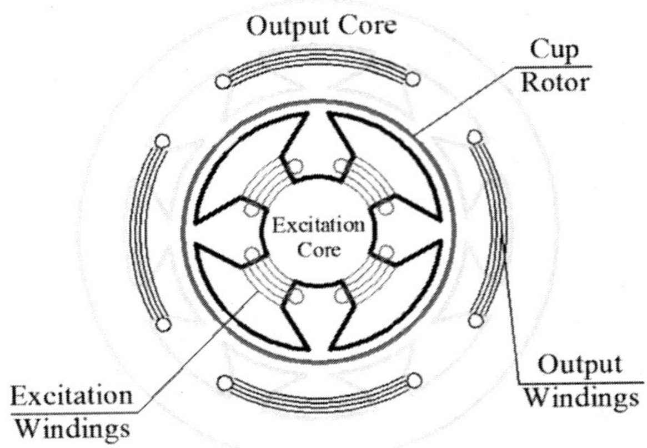

Figure 3. Sectional sketch view of the sensor.

2.2. The Principle of Sensors

The magnetic circuit is shown in Figure 4. When the excitation windings have DC current I_1 assuming the magnetic circuit is symmetrical, the magnetic flux Φ_1 of the magnetic circuit can be expressed by the following equation according to the magnetic circuit theorem [19]:

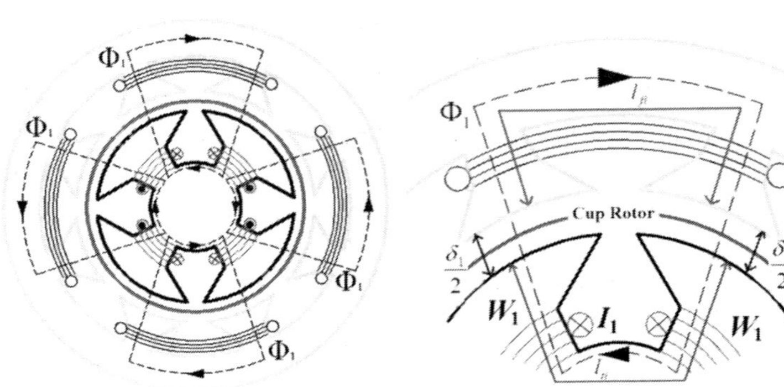

Figure 4. The magnetic circuit of Φ_1.

$$\Phi_1 = \frac{F_1}{R_{m1}} = \frac{2W_1 I_1}{\frac{\delta_1}{\mu_0 S_1} + \frac{l_R}{\mu_R S_{R1}} + \sum_{i=1}^{n} \left(\frac{l_{fi}}{\mu_f S_{fi}} + \frac{l_{ti}}{\mu_t S_{ti}} \right)} = K_1 I_1 \quad (1)$$

In which:

W_1 The effective windings number of the excitation windings

δ_1 The effective length through the air-gap Φ_1

S_1 The effective area through the air-gap Φ_1

μ_0 The magnetic permeability of the air

l_R The effective thickness of the cup-shaped rotor

μ_R The magnetic permeability of the cup-shaped rotor

S_{R1} The effective area through the cup-shaped rotor Φ_1

l_{fi} The each segment effective length through the outer stator core of magnetic flux Φ_1

S_{fi} The each segment effective area through the outer stator core of magnetic flux Φ_1

μ_f The magnetic permeability of the outer stator core

l_{ti} The each segment effective length through the inner stator core of magnetic flux Φ_1

S_{ti} The each segment effective area through the inner stator core of magnetic flux Φ_1

μ_i The magnetic permeability of the inner stator core

$$K_1 = \frac{2W_1}{\frac{\delta_1}{\mu_0 S_1} + \frac{l_R}{\mu_R S_{R1}} + \sum_{i=1}^{n} \frac{l_{fi}}{\mu_f S_{fi}} + \frac{l_{ti}}{\mu_t S_{ti}}}$$

The cup-shaped rotor can be seen as a number of squirrel-cage rotors. When the magnetic flux is cut counterclockwise by the cup-shaped rotor with a speed of n (unit is r/min), the generated motional electromotive force is:

$$e_R = C_e \Phi_1 n \tag{2}$$

where C_e is the structure constant associated with the cup-shaped rotor, $C_e = \frac{\pi D}{60\tau}$, where τ is the pole moment of the stator and D is the outer diameter of the cup-shaped rotor. The resistance of the cup-shaped rotor is r_R, and the reactance of cup-shaped rotor can be ignored, the current i_R in the cup-shaped rotor is:

$$i_R = \frac{e_R}{r_R} \tag{3}$$

i_R generates a magnetic field, and its magnetic circuit is shown in Figure 5.

Similarly, the magnetic flux Φ_R generated by i_R can be expressed by the following equation according to the magnetic circuit theorem:

$$\Phi_R = \frac{W_R i_R}{\frac{\delta_R}{\mu_0 S_R} + \frac{l_R}{\mu_R S_{R2}} + \sum_{i=1}^{n}\left(\frac{l_{Fi}}{\mu_f S_{Fi}} + \frac{l_{Ti}}{\mu_t S_{Ti}}\right)} = K_R i_R \tag{4}$$

In which:

W_R The equivalent number of windings for the cup-shaped rotor

δ_R The effective length through the air-gap Φ_R

S_R The effective area through the air-gap Φ_R

S_{R2} The effective area through the cup-shaped rotor of the magnetic flux Φ_R

l_{Fi} The each segment effective length of magnetic flux Φ_R through the outer stator core

S_{Fi} The each segment effective area of magnetic flux Φ_R through the outer stator core

l_{Ti} The each segment effective length of magnetic flux Φ_R through the stator core

S_{Ti} The each segment effective area of magnetic flux Φ_R through the stator core

$$K_R = \frac{W_R}{\frac{\delta_R}{\mu_0 S_R} + \frac{l_R}{\mu_R S_{R2}} + \sum_{i=1}^{n} \frac{l_{Fi}}{\mu_f S_{Fi}} + \frac{l_{Ti}}{\mu_t S_{Ti}}}$$

The total effective windings of the output windings are W_2. According to the principle of electromagnetic induction, the induction electromotive force in the output windings induced by the magnetic flux Φ_R is exhibited as the following equation:

$$e_o = -W_2 \frac{d(2\Phi_R)}{dt} \quad (5)$$

By the combination of Equations (1–5), the following equation can be obtained:

$$e_o = \frac{-2K_1 K_R C_e W_2 W_R I_1}{r_R} \frac{dn}{dt} \quad (6)$$

From the above equation, we can find that the induction electromotive force of output windings e_o is proportional to the rotation angular acceleration of the cup-shaped rotor.

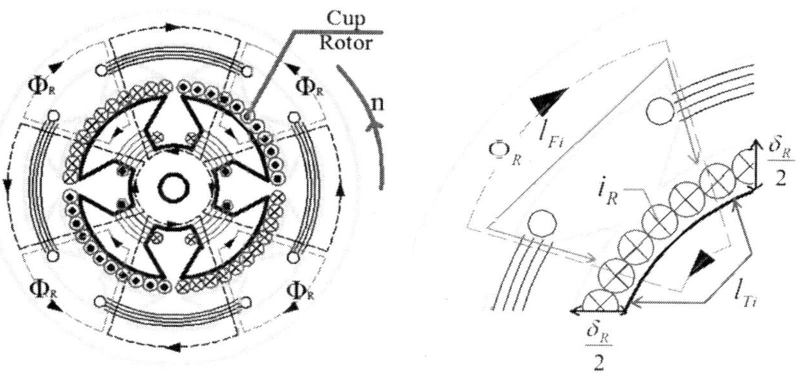

Figure 5. The magnetic circuit of Φ_R.

3. THE MATHEMATICAL MODEL OF SENSOR

3.1. Transfer Function

The equivalent circuit of the working sensor is shown in Figure 6:

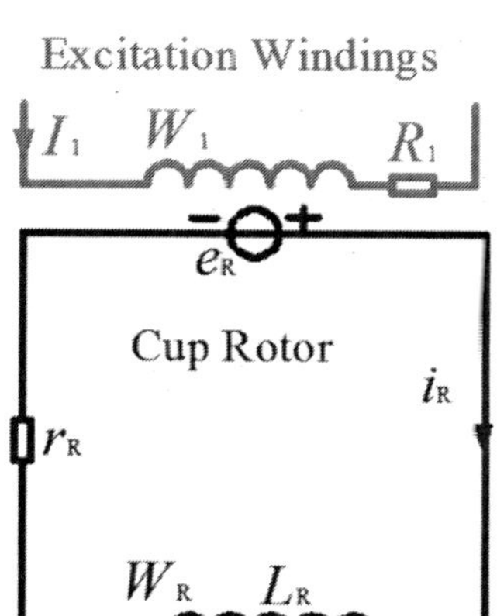

Figure 6. The equivalent circuit of the working sensor.

In Figure 6:

R_1 The resistance of excitation windings

e_R The motional electromotive force of cup-shaped rotor

L_R The inductance of the cup-shaped rotor

r_R The resistance of cup-shaped rotor

i_R The current of cup-shaped rotor

e_o The induction electromotive force of output windings

i_2 The current of output windings

L_x The leakage reactance of output windings

r_2 The resistance of output windings

Z The external load of output windings

For the cup-shaped rotor, the voltage balance equation is:

$$e_R = i_R r_R + L_R \frac{di_R}{dt} \tag{7}$$

For the output windings, the voltage balance equation is:

$$e_o = i_2 r_2 + L_x \frac{di_2}{dt} + i_2 Z \tag{8}$$

The output voltage of sensor, namely the voltage across load is:

$$u_2 = i_2 Z \tag{9}$$

The Laplace transform is demonstrated for the Equations (5–9), after combination, the transfer function of sensor is obtained:

$$G(s) = \frac{L[u_2]}{L[\frac{dn}{dt}])} = \frac{u_2(s)}{sn(s)} = \frac{-C_e K_1 K_R W_2 Z I_1}{(r_R + L_R s)(r_2 + Z + L_x s)} \tag{10}$$

It is seen that the intrinsic oscillation frequency (cutoff frequency) is $\omega_n = \sqrt{\frac{r_R(r_2 + Z)}{L_R L_x}}$ and the damping ratio is $\xi = \frac{(r_2 + Z)L_R + r_R L_x}{2\sqrt{(r_2 + Z)L_R \cdot r_R L_x}}$ for the working sensor system, respectively. The poles of the system are $s_1 = -\frac{r_R}{L_R}$ and $s_2 = -\frac{r_2 + Z}{L_x}$ respectively. Owing to the fact that the load resistance is large and the output windings leakage inductance L_x is small, the dominant apex of the system is s_1, that is $G(s) \approx -\frac{2C_e K_1 K_R W_2 Z I_1}{r_R + L_R s}$. The system can be approximated as a series of the proportional components and the inertial element, and the time constant of the inertial element is L_R/r_R. High resistivity material can be used to prepare the cup-shaped rotor, in order to improve the response speed of the system.

3.2. State-Space Model

Appropriate state variables are selected, the diagonal form of the system state-space model can be obtained through the transfer function of the system (10):

$$\begin{aligned}\begin{bmatrix}\dot{x}_1\\ \dot{x}_2\end{bmatrix} &= \begin{bmatrix} -\frac{r_R}{L_R} & 0 \\ 0 & -\frac{r_2+Z}{L_x}\end{bmatrix}\begin{bmatrix}x_1\\ x_2\end{bmatrix} + \begin{bmatrix}1\\ 1\end{bmatrix}\frac{dn}{dt} = \mathbf{A}\begin{bmatrix}x_1\\ x_2\end{bmatrix}+\mathbf{B}\frac{dn}{dt} \\ y &= \left[\frac{2C_eK_1K_RW_2ZI_1L_R}{r_RL_x-(r_2+Z)L_R} \quad \frac{2C_eK_1K_RW_2ZI_1L_x}{(r_2+Z)L_R-r_RL_x}\right]\begin{bmatrix}x_1\\ x_2\end{bmatrix} = \mathbf{C}\begin{bmatrix}x_1\\ x_2\end{bmatrix}\end{aligned} \quad (11)$$

The corresponding state variables scheme is shown in Figure 7:

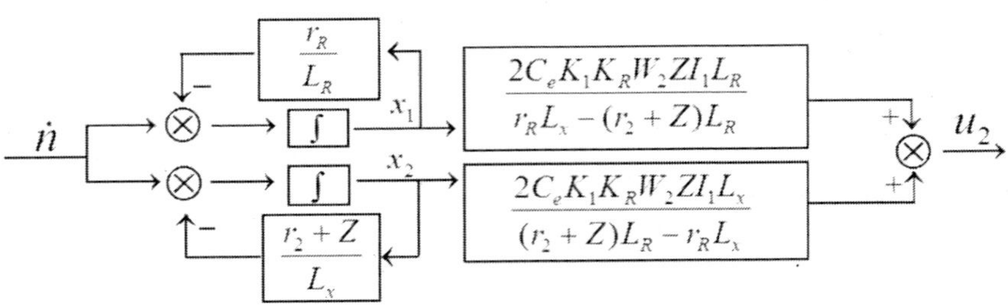

Figure 7. The state variables scheme of sensor.

The stability of the system can be divided into external stability and internal stability. The external stability is defined as the influence of the external input on the response of the system at the initial state, which can be judged according to the equation root situation of the closed-loop characteristic equation for the system. The internal stability is only related to the impact of the system itself at the initial state, also called Lyapunov asymptotical stability.

Assuming $\mathbf{P} = \begin{bmatrix} p_1 & p_2 \\ p_3 & p_4 \end{bmatrix}$, the state matrix is introduced in Lyapunov equation, the solution is:

$$\begin{cases} p_1 = \dfrac{L_R}{2r_R} \\ p_2 = p_3 = 0 \\ p_4 = \dfrac{L_x}{2(r_2 + Z)} \end{cases} \tag{12}$$

Owing to $p_1 > 0$ and $\begin{vmatrix} p_1 & p_2 \\ p_3 & p_4 \end{vmatrix} = \dfrac{L_R L_x}{4(r_2 + Z)r_R} > 0$, the matrix P is positive definite, so the working sensor is Lyapunov asymptotic stable over a wide range.

4. THE SENSOR CALIBRATION

4.1. Composition of the Calibration Equipment

The calibration equipment of the sensor is shown in Figure 8, which is composed of bracket ①, angular acceleration sensor ②, couplings ③, mass ④, coil spring ⑤, torsion bar ⑥, base ⑦, angle sensor ⑧, digital oscilloscope ⑨, and DC power ⑩.

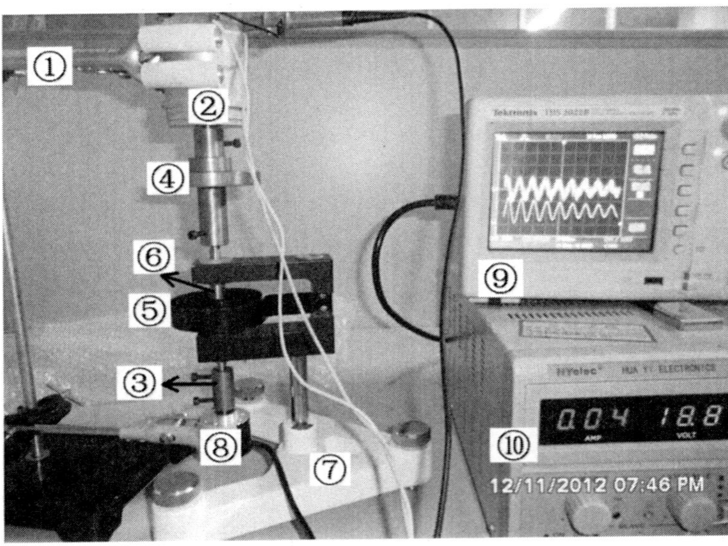

Figure 8. The calibration equipment of angular acceleration sensor.

The bracket is used to fix angular acceleration sensor ② and high-precision angle sensor ⑧; The coupling ③ is used for the coaxial connection of the angular acceleration sensor ② and high-precision angle sensor ⑧; The mass ④ is a self-made iron hoop with holes and has variable quality for selection and combination. The moment of inertia for the calibration system can be changed to realize calibration at different frequencies through selecting and combining masses; the combination of a large mass and a small mass is shown in the Figure; the coil spring ⑤, the torsion bar ⑥ and the base ⑦ form a torsional pendulum, which is the power source of torsional vibration for the entire system. The torsion bar ⑥ passes the mass and connects with the angular acceleration sensor ② through the upper end of the coupling ④, and its bottom end is connected with the end of the angle sensor ⑧; The dual channel input of the digital oscilloscope ⑨ connects with the angular acceleration sensor and the angle sensor, respectively, for the acquisition and preservation of the output waveform of the two sensors during the experiments.

4.2. The Principle of the Calibration Equipment

An initial force is applied to the torsion bar ⑥, resulting in the coil spring ⑤ rotating an angle θ to produce the restoring moment $T_{1s} = k\theta$, in which k is the stiffness coefficient of the spring; The system begins to rotate under this restoring moment when the initial force disappears. In the rotation process, the system is affected by the air resistance torque $T_{2s} = -\gamma \frac{d\theta}{dt}$, in which γ is the drag torque coefficient; and also affected by the bearing friction resistance torque $T_{3s} = -\alpha \frac{d\theta}{dt}$, in which α is the friction torque coefficient; Assuming the system moment of inertia is J, we can obtain the following equation according to the torque balance equation:

$$J\frac{d^2\theta}{dt^2} + \alpha\frac{d\theta}{dt} + \gamma\frac{d\theta}{dt} + k\theta = 0 \tag{13}$$

Assuming $2\beta = \frac{\gamma + \alpha}{J}$ and $\omega_n^2 = \frac{k}{J}$, where β is the damping coefficient of the system, ω_n is the intrinsic oscillation angular frequency of the system:

$$\frac{d^2\theta}{dt^2} + 2\beta\frac{d\theta}{dt} + \omega_n^2\theta = 0 \tag{14}$$

The solution for above equation is $\theta(t) = \frac{\theta_0}{\sqrt{1-\xi^2}} e^{-\beta t}\cos(\omega_d t - \varphi)$, in which θ_0 is the initial amplitude of the pivot angle, $\omega_d = \omega_n\sqrt{1-\xi^2}$ is the damped oscillation angular frequency, damping ratio is $\xi = \frac{\gamma}{2\sqrt{kJ}} = \frac{\beta}{\omega_n}$. It is found that the vibration amplitude decays with time; the bigger the damping factor is, the faster it will decay. Assuming T is the cycle damping vibration for the pendulum, $\frac{\theta_0}{\theta_N} = e^{\beta NT}$ can be obtained if the initial amplitude θ_0 and the amplitude θ_N of the N-th cycle, as well as the used time of N cycles are measured. The damping coefficient can be expressed as:

$$\beta = \frac{1}{NT}\ln\frac{\theta_0}{\theta_N} \tag{15}$$

The rotation cycle of the system is $T = \frac{2\pi}{\omega_d} = \frac{2\pi}{\omega_n\sqrt{1-\xi^2}} = \frac{2\pi}{\sqrt{\omega_n^2 - \beta^2}}$, then $\frac{k}{J} = \frac{4\pi^2}{T^2} + \beta^2$ can be inferred. The angular rate of the pendulum is $\frac{d\theta}{dt} = 0$ when system at the amplitude position. At this time, Equation (13) can be simplified as:

$$\frac{d^2\theta}{dt^2} = -\frac{k}{J}\theta = -(\frac{4\pi^2}{T^2} + \beta^2)\theta \tag{16}$$

where β is calculated by Equation (15) through experiments, and then the cycle T and the rotor angle θ are obtained through the output waveform of angle sensor saved by the oscilloscope. From Equation (16), we can get the actual angular acceleration value of the system, corresponding to the output voltage of the angular acceleration sensor. Additionally, we can also obtain the sensitivity coefficient of the angular acceleration sensor.

4.3. The Calibration Experiment Results

The DC resistance of the excitation windings for the angular acceleration sensor is 838 Ω. After a long-time test at the working temperature, excitation windings voltage U_1 = 40 V is selected for the angular acceleration sensor. In this case, the temperature of the sensor is about 25 to 28 Celsius degrees; the initial angle is about 180° when the external force is applied on the coil spring, the output signals of angular acceleration sensor and the angle sensor are shown in Figure 9 and Figure 10 after the external force disappears. The output electrical signal of the angular acceleration sensor is de-noised by a Rigrsure wavelet; the obtained signal is shown in the Figure 11. The calculated results are shown in Table 1 by selecting the signal peaks of angular acceleration sensor and the angle sensor, according to the relevant formulas.

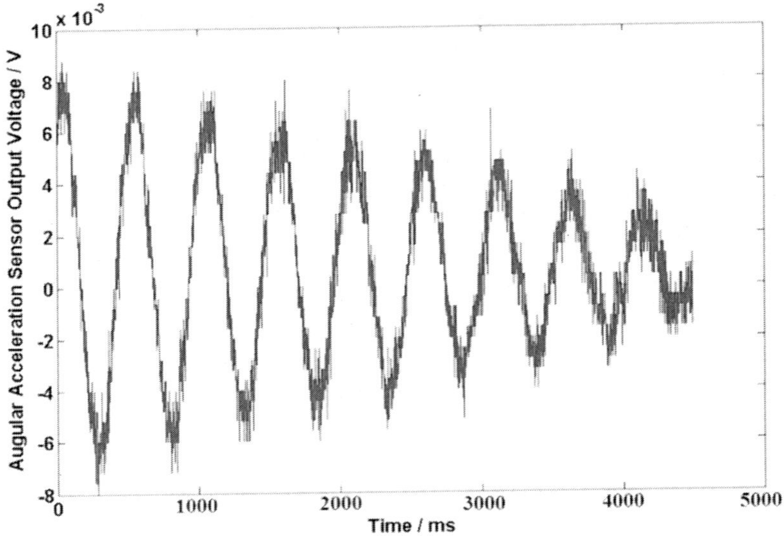

Figure 9. Output signals of angular acceleration sensor.

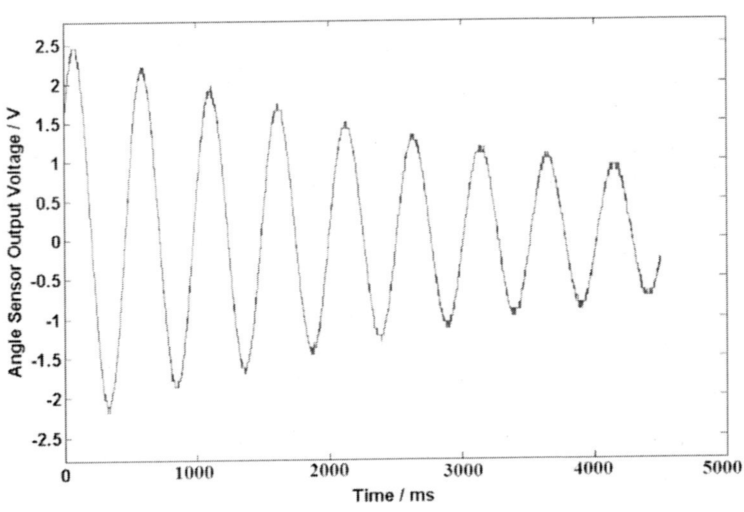

Figure 10. Output signals of angle sensor.

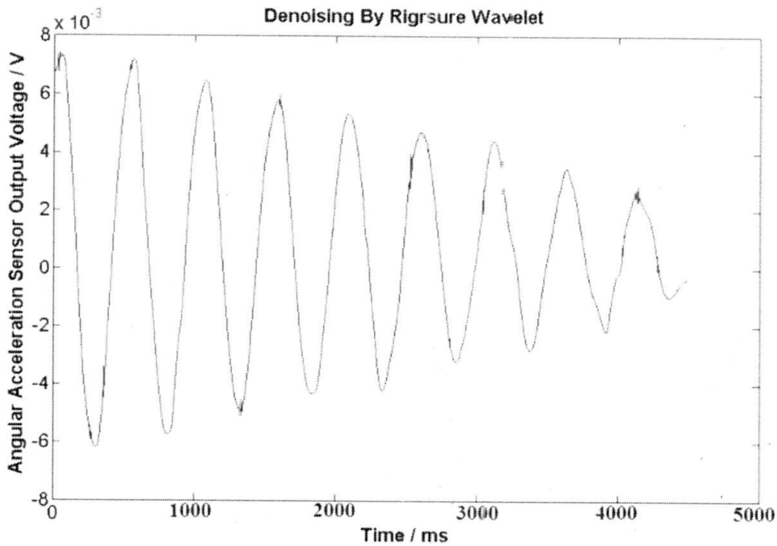

Figure 11. The de-noised signals by Rigrsure wavelet.

Table 1. Experimental calibration data.

The peak voltage of the angular acceleration sensor (mV)	7.32	7.15	6.44	5.44	5.06	4.48	4.04	3.61	3.09
The peak voltage of angle sensor (V)	2.30	2.24	1.90	1.66	1.50	1.32	1.18	1.08	0.94
The time corresponding peak (s)	0.071	0.586	1.098	1.606	2.12	2.626	3.138	3.656	4.158
Oscillation period T/(s)	(0.515 + 0.512 + 0.508 + 0.494 + 0.506 + 0.512 + 0.518 + 0.502)/8 = 0.5084								
Angle values (rad)	2.89	2.815	2.388	2.086	1.885	1.659	1.483	1.357	1.181
System damping ratio coefficient β	$[\ln(3.091 \div 1.181)] \div 8 \div T \approx 0.2365$								
Angular acceleration value (rad/s^2)	−441.5	−430	−364.8	−318.6	−287.9	−253.4	−226.5	−207.3	−180.4
Sensitivity coefficient (mv/Krad/s^2)	16.58	16.63	17.65	17.08	17.58	17.67	17.83	17.37	17.23

Matlab is used to fit the data in Table 1 and the obtained results are shown in Figure 12. It is found that the sensitivity coefficient of the angular acceleration sensor is about 17.29 mv/(Krad/s^2).

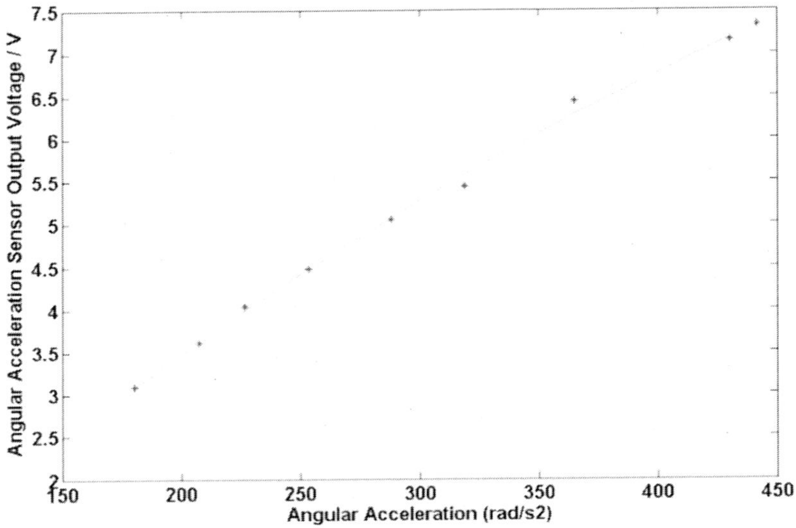

Figure 12. Calibration curve of angular acceleration sensor.

4.4. Analysis of the Calibration Experiments

There will exist a phase difference when the angular acceleration sensor and the angle sensor are working in practice. From the calibration principle, we can see that the peak value of the angular acceleration is the corresponding amplitude position of the pendulum. Therefore, the peak voltages of the two sensors are needed in the calibration process and the phase difference shows little impact on the calibration results. Additionally, the aerostatic bearing can be considered to reduce the influence of bearing friction on calibration results; the calibration range of the system is limited and only suitable for the calibration of the angular acceleration in the lower frequency owing to the constraint of the mechanical structure and the limitation of variation range, even though the calibration frequency can be changed by the mass to a certain extent; the obtained calibration frequency is not continuous because the calibration system depends on the mass.

In order to solve the issues of non-continuous calibration point and lower frequency band, an auxiliary power source can be added to the electromagnetic force to realize a forced vibration state when the calibration equipment is working. The linear regulator of the vibration frequency can be realized by changing the rotational speed of the power source, resulting in the extension of the calibration frequency. These improvement ideas will be demonstrated in our future experiments.

5. ERROR ANALYSIS

5.1. Fluctuations of Excitation Voltage

From the sensor principle, it is recognized that magnetic flux Φ_1 generated by the excitation current includes the variation component $\Delta\Phi_1$ if the excitation voltage has fluctuations. The cup-shaped rotor cuts the magnetic flux Φ_1 with a constant rotation speed, so its motional electromotive force e_R includes the variation component Δe_R, resulting in the existence of a current change component Δi_R for the cup-shaped rotor, and the generated magnetic flux Φ_R also contains a variation component $\Delta\Phi_R$. At this time, output windings will generate the corresponding induced electromotive force, known as the residual voltage. In order to overcome the abovementioned disadvantages of the sensor, a constant current source or a permanent magnet can be used to replace the existing excitation circuit to guarantee the constancy of the magnetic flux Φ_1.

5.2. Effect of Temperature

The properties of the excitation windings, the output windings, the resistance of the cup-shaped rotor and the magnetic material may change because of the change in ambient temperature and the heating from extended use of the sensor. These property changes show influences on the output induced electromotive force of the sensor, resulting in the instability of the output performance. The character of the temperature-sensitivity coefficient of the sensor is shown in Figure 13. In this experiment, DS18B20 is used to detect the temperature of the sensor case. According to the results, the sensor sensitivity coefficient will decrease slightly with increasing temperature.

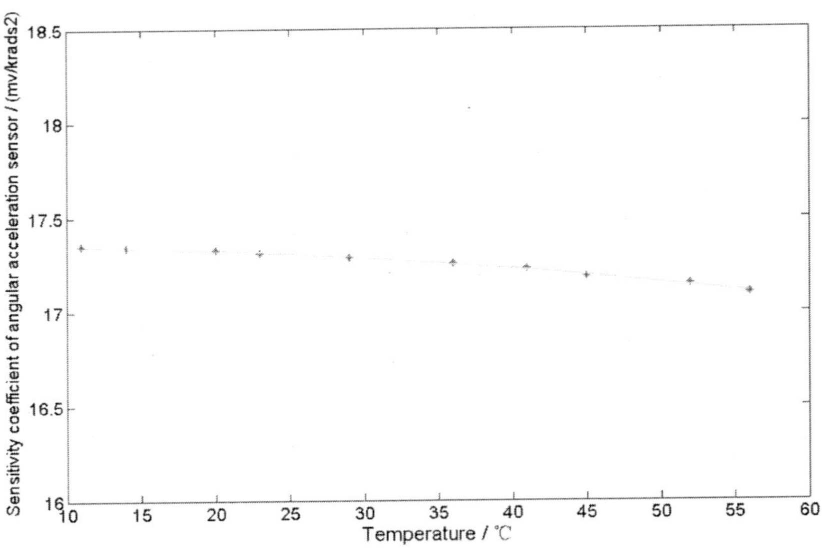

Figure 13. Temperature-sensitivity coefficient characteristics of the sensor.

In order to decrease the impact of temperature, temperature compensation should be applied. A negative temperature coefficient thermistor can be concatenated in the excitation and the output circuit to compensate the temperature variation influence.

5.3. Nonlinearity Errors

The cup-shaped rotor of actual sensor is a first-order circuit composed of a resistance and inductance, so the change of the current i_R lags induces an electromotive force e_R due to the influence of the inductance L_R, resulting in the distortion of the symmetrical magnetic circuit: the magnetic flux Φ_R generated by the current i_R varies with the change of rotational speed, and the change component of Φ_R demagnetizes magnetic flux Φ_1, causing the change of the magnetic flux Φ_1. As a result, linear relationships of the output potential and the angular acceleration are destroyed, resulting in the presence of a non-linear error.

6. CONCLUSIONS

The angular acceleration sensor presented in this paper shows strong practicality without limitation on the rotation angle range. It only needs to connect with the measured rotating system coaxially when working. The structure and principle of the calibration equipment are described, and the calibration curves and performance indicators of the sensor are obtained by experiments. In our following work efforts will be made to examine the practical applications for this sensor in view of the extensive applications of angular acceleration sensors, such as transmission noise measurement and fault diagnosis for complex gearboxes or the internal combustion engine, fuel injection timing and fuel injection amount control for automotive engines, as well as the kinetic analysis for mechanical arm joints and so on.

ACKNOWLEDGMENTS

The work was sponsored in part by the Jiaxing Science and Technology Research Project of China under Grant (2012AY1021) and in part by the Zhejiang Provincial Department of Education Scientific Research Project of China under Grant (Y201226082) and in part by the Zhejiang Province Major Science and Technology Project of China under Grant (2008C01011-4) and in part by the Jiaxing University Self-made Experimental Teaching Instrument and Equipment Research Project (ZS201202) and in part by the Jiaxing University Nanhu College Key Scientific Research Project.

CONFLICT OF INTEREST

The authors declare no conflict of interest.

REFERENCES

1 Aramaki, Y.; Nozaki, D.; Masani, K.; Sato, T.; Nakazawa, K.; Yano, H. Reciprocal angular acceleration of the ankle and hip joints during quiet standing in humans. *Exp. Brain Res.* **2011**, *136*, 463–473.

2 Song, Y.; Zhang, P.; Obcemea, C.; Mueller, B.; Chandra, C.; Mychalczak, B. Dosimetric Effects of Gantry Angular

Acceleration and Deceleration in Volumetric Modulated Radiation Therapy. IFMBE Proceedings of the World Congress on Medical Physics and Biomedical Engineering, Munich, Germany, 7–12 September 2009; Volume 25, pp. 1046–1050.

3 Kimpara, H.; Iwamoto, M. Mild traumatic brain injury predictors based on angular accelerations during impacts. *Ann. Biomed. Eng.* **2012**, *40*, 114–126.

4 Angeles, J. The angular-acceleration tensor of rigid-body kinematics and its properties. *Arch. Appl. Mech.* **1999**, *69*, 204–214.

5 Yoganandan, N.; Pintar, F.A.; Zhang, J.Y.; Stemper, B.D.; Philippens, M. Upper neck forces and moments and cranial angular accelerations in lateral impact. *Ann. Biomed. Eng.* **2008**, *36*, 406–414.

6 Chan, C.-K.; Lo, S.-C.; Huang, Y.-C.; Wu, M.C.; Wang, M.-Y.; Fang, W.L. Poly-si based two-axis differential capacitive-sensing accelerometer. *IEEE Sens. J.* **2012**, *12*, 3301–3308.

7 Partridge, A.; Reynolds, J.K.; Chui, B.W.; Chow, E.M.; Fitzgerald, A.M.; Zhang, L.; Maluf, N.I.; Kenny, T.W. A high-performance planar piezo-resistive accelerometer. *J. Micro Electro Mech. Syst.* **2000**, *9*, 58–66.

8 Qu, H.; Fang, D.; Xie, H. A Single-Crystal Silicon 3-Axis CMOSMEMS Accelerometer. Proceedings of the IEEE Sensors, Vienna, Austria, 24–27 October 2004; pp. 661–664.

9 Sun, C.-M.; Tsai, M.-H.; Liu, Y.-C.; Fang, W. Implementation of a monolithic single proof-mass tri-axis accelerometer using CMOSMEMS technique. *IEEE Trans. Electron. Devices* **2010**, *57*, 1670–1679.

10 Ovaska, S.J.; Valiviita, S. Angular acceleration measurement: A review. *IEEE Trans. Instrum. Measur.* **1998**, *47*, 1211–1217.

11 Vol Moody, M.; Paik, H.J.; Canavan, E.R. Principle and performance of a superconducting angular accelerometer. *Rev. Sci. Instrument.* **2003**, *74*, 1310–1318.

12 Sparks, D.R.; Huang, X.; Higdon, W.; Johnson, J.D. Angular rate sensor and accelerometer combined on the same micro-machined CMOS chip. *Micro-Syst. Technol.* **1998**, *4*, 139–142.

13 Li, J.L.; Fang, J.C.; Du, M.; Dong, H.F. Analysis and fabrication of a novel MEMS pendulum angular accelerometer with electrostatic actuator feedback. *Micro-Syst. Technol.* **2012**, *19*, 9–16.

14 Tomikawa, Y.; Okada, S. Piezoelectric Angular Acceleration Sensor. Proceedings of the IEEE Ultrasonic Symposium, Honolulu, HI, USA, 5–8 October 2003; pp. 1346–1349.

15 Yamamoto, Y.; Jikuya, K.; Nakamura, T.; Kusuhara, T.; Michinishi, H.; Okamoto, T. Precise Measurement System for Knee Joint Motions at Pendulum Test Using Accelerometer-Detection of Angular Acceleration by Two Linear Accelerometers. International Federation for Medical and Biological Engineering Proceedings of the World Congress on Medical Physics and Biomedical Engineering, Munich, Germany, 7–12 September 2009; Volume 25, pp. 165–168.

16 Mizuno, J.; Nottmeyer, K.; Amemori, M.; Kanai, Y.; Kobayashi, T. The study of silicon bulk micro-machined angular acceleration sensor. *JSAE Rev.* **2000**, *21*, 79–84.

17 Marat-Mendes, R.; Dias, C.J.; Marat-Mendes, J.N. Measurement of the angular acceleration using a PVDF and a piezo-composite. *Sens. Actuators B-Phys.* **1999**, *76*, 310–313.

18 Wolfaardt, H.J.; Heyns, P.S. Dynamic modeling of a novel micro-fluidic channel angular accelerometer. *J. Vibr. Control* **2008**, *14*, 451–467.

19 Tang, Y.M. *Electric Machinery*; China Machine Press: Beijing, China, 2011; pp. 13–16.

Index

P

R

S

T